Sensory Evaluation Techniques

Volume II

Authors

Morten Meilgaard, D. Tech.
Vice President, Research
The Stroh Brewery Company
Detroit, Michigan

Gail Vance Civille, B.S.
Consultant
Sensory Spectrum
East Hanover, New Jersey

B. Thomas Carr, M.S.
Senior Statistician
The NutraSweet Co., R & D
Mt. Prospect, Illinois

CRC Press
Taylor & Francis Group
Boca Raton London New York

CRC Press is an imprint of the
Taylor & Francis Group, an **informa** business

CRC Press
Taylor & Francis Group
6000 Broken Sound Parkway NW, Suite 300
Boca Raton, FL 33487-2742

Reissued 2019 by CRC Press

© 1987 by Taylor & Francis Group, LLC
CRC Press is an imprint of Taylor & Francis Group, an Informa business

No claim to original U.S. Government works

A Library of Congress record exists under LC control number:

Publisher's Note
The publisher has gone to great lengths to ensure the quality of this reprint but points out that some imperfections in the original copies may be apparent.

Disclaimer
The publisher has made every effort to trace copyright holders and welcomes correspondence from those they have been unable to contact.

ISBN 13: 978-0-367-26177-1 (hbk)
ISBN 13: 978-0-367-26178-8 (pbk)
ISBN 13: 978-0-429-29185-2 (ebk)

Visit the Taylor & Francis Web site at http://www.taylorandfrancis.com and the
CRC Press Web site at http://www.crcpress.com

To Manon, Frank and Cathy

PREFACE

How does one plan, execute, complete, analyze, interpret, and report sensory tests? Hopefully, the practices and recommendations in these two volumes, cover all of those phases of sensory evaluation. The test is meant to provide enough information for a food technologist, a research and development scientist, a cereal chemist, a perfumer, or a similar professional working in industry, academics, or government to conduct good sensory evaluation. The books should also supply useful background to market research, advertising, and legal professionals who need to understand the results of sensory evaluation. They could also give a sophisticated general reader the same understanding.

As a "how to" for professionals, the text aims at a clear and concise presentation of practical solutions, accepted methods, and standard practices. The authors at first intended not to devote text and readers' time to resolving controversial issues. Unfortunately, we encountered quite a few which had to be tackled. This is the first book to give an adequate solution to the subject of similarity testing, see Chapter 6, Section II.G and Statistical Tables T11, T12, and T13 at the end of Volume II. Fully half of all sensory tests are done for purposes of similarity testing, for example when an ingredient must be substituted for another which has become unavailable or too expensive, or when a change in processing is caused by replacement of an old or inefficient piece of equipment. Another first is the unified statistical treatment of all ranking tests with the Friedman statistic, in preference to Kramer's tables. We have taken a fresh look at all statistical methods used for sensory tests and hope that you like our straightforward approach.

Also new is a system called Spectrum®, developed by one of us (GVC) for designing procedures of descriptive analysis (Chapter 8). The philosophy behind Spectrum is twofold; (1) to tailor the test to suit the objective of the study (and not to suit a prescribed format) and (2) that the choice of terminology and reference standards are factors too important to be left to the panelists, however well trained. These items should be chosen by the sensory analyst who needs all the accumulated experience of his or her profession for the task.

The authors wish the book to be cohesive and readable; we have tried to substantiate our directions and organize each section so as to be meaningful. We do not want the book to be a turgid set of tables, lists, and figures. We hope to have provided structure to the methods, reason to the procedures, and coherence to the outcomes. We want this to be a reference text that can be read for understanding as well as a handbook that can serve to summarize sensory evaluation practices.

The organization of the chapters and sections is also straightforward. Chapter 1 lists the steps involved in a sensory evaluation project and Chapter 2 briefly reviews the workings of our senses. In Chapter 3, we list what's required of the equipment, the tasters, and the samples, while in Chapter 4, we have collected a list of those psychological pitfalls which invalidate many otherwise good studies. Chapter 5 discusses how sensory responses can be measured in quantitative terms. Chapter 6 lists all the common sensory tests for difference, the triangle, duo-trio, etc., as well as the various attribute tests in use, such as ranking and numerical intensity scaling. Thresholds and just-noticeable differences are briefly discussed in Chapter 7, followed by what we consider the main chapters, Chapter 8 on descriptive testing, Chapter 9 on affective tests (consumer tests), and Chapter 10 on selection and training of tasters.

The body of text on statistical procedures is found in Chapters 11 and 12 but in addition, each method (triangle, duo-trio, etc.) is followed by a number of examples showing how statistics are used in the interpretation of each. Basic statistical concepts such as null and alternative hypotheses, Type I and Type II errors and their relation to alpha, beta, and the sample size "n", one-sided vs. two-sided tests, etc. are presented in Chapter 11. We refrain from detailed discussion of statistical theory, preferring instead to give examples. Included in Chapter 12 are discussions of some commonly used experimental designs, such as the randomized block, split plot, and balanced incomplete block. Chapter 12 also includes a discussion of multiway treatment structures, such as factorial experiments and the ever more frequently used statistical technique of Response Surface Methodology (RSM) in which predictive equations are developed that relate a set of sensory responses to the levels of a set of processing parameters or, alternatively to the proportions of a set of ingredients. Also in Chapter 12 the use of multivariate techniques is briefly discussed. This is a subject still in its infancy and future editions of this book probably will contain expanded treatment of this topic.

At the end of Volume II, the reader will find guidelines for the choice of techniques and for reporting results, plus the usual glossaries, indexes, and statistical tables.

With regard to terminology, the terms "subject", "panelist", "judge", "respondent", and "assesor" are used interchangeably, as are "he", "she", and "(s)he" for the sensory analyst (the sensory professional, the panel leader), and for individual panel members.

Morton Meilgaard
Gail Vance Civille
B. Thomas Carr

ACKNOWLEDGMENTS

Thanks are due to our associates at work and our families at home for thoughts and ideas, and for material assistance with typing and editing. Numerous individuals and companies have contributed to the ideas of this book and their application in practice. In particular we wish to acknowledge the emotional contributions of Manon, Frank and Cathy, and the technical ones of our colleagues in ASTM's Subcommittee E-18 on Sensory Evaluation, of whom we would like to single out Louise Aust, Donna Carlton, Andrew Dravnieks, Jean Eggert, Patricia Prell, John J. Powers, and Elaine Skinner. Thanks are due to Rose Marie Pangborn and Ann Noble at Davis, Calif., to Elizabeth Larmond at Agriculture Canada, and to Roland Harper and Derek Land in the U.K. for suggestions and discussions over the years, and to Erik Knudson, Stephen Goodfellow, Dan Grabowski, Cathy Foley, and Clare Dus for help with illustrations, layout, and ideas, also to the Stroh Brewery Co. and Searle Inc. for permission to publish and for the use of facilities and equipment.

THE AUTHORS

Morten C. Meilgaard, M.Sc., D. Tech., F. I. Brew. is currently Vice President, Research at the Stroh Brewery Company in Detroit, Mich. He graduated in 1951 in biochemistry and engineering from the Technical University of Denmark, then returned in 1982 to receive a doctorate for a dissertation on beer flavor compounds and their interactions. After 6 years as a chemist at the Carlsberg Breweries, he worked from 1957—1967 as a worldwide consultant on beer and brewing. He then served for 6 years as Director of Research for Cerveceria Cuauhtemoc in Monterrey, Mexico before taking up his present employment in 1973.

Dr. Meilgaard's professional interest is the biochemical and physiological basis of flavor, and more specifically the flavor compounds of hops and beer and the methods by which they can be identified, namely chemical analysis coupled with sensory evaluation techniques. He has published over 50 papers and received the Schwarz Award for studies of compounds that affect beer flavor. He is founder and past president of the Hop Research Council of the U.S.A., also chairman of the Scientific Advisory Committee of the U.S. Brewers Association, and for 14 years he has been the chairman of the Subcommittee on Sensory Analysis of the American Society of Brewing Chemists.

Gail Vance Civille is the president of Sensory Spectrum, Inc., a consulting firm in the field of sensory evaluation of foods, beverages, pharmaceuticals, paper, fabrics, personal care and other consumer products. She is involved in the selection, implementation and analysis of test methods for solving problems in quality control, research, development, production and marketing. She has trained several flavor and texture descriptive profile panels in her work with industry, universities and government.

As a Course Director for the Center for Professional Advancement, she has conducted several workshops and courses in basic sensory evaluation methods as well as advanced methods and theory. In addition, she has been invited to speak to several professional organizations on different facets of sensory evaluation.

Ms. Civille has published several articles on general sensory methods, as well as sophisiticated descriptive flavor and texture techniques. After graduating from the College of Mount Saint Vincent, New York with a B.S. degree in Chemistry, Ms. Civille began her career as a product evaluation analyst with the General Foods Corporation.

B. Thomas Carr is Senior Statistician in the NutraSweet Company, Research & Development Division, where he works closely with NutraSweet's Sensory Evaluation Group on the design and analysis of a wide variety of sensory studies in support of Product/Process Development, QA/QC, and Research Guidance Consumer Tests. Mr. Carr is also actively involved in implementing NutraSweet's automated sensory data-acquisition and data-handling system. Previously Mr. Carr was Supervisor of Statistical Services for CPC International/Best Foods.

Mr. Carr is actively involved in statistical training of scientists, both within NutraSweet and, internationally, in collaboration with the second author, Ms. Civille. In addition, he has been an invited speaker to several professional organizations on the topics of statistical methods and statistical consulting in industry.

Since 1979 Mr. Carr has supported the development of new food ingredients, consumer food products, and OTC drugs by integrating the statistical and sensory evaluation functions into the mainstream of the product development effort. This has been accomplished through the application of a wide variety of statistical techniques including design of experiments, response surface methodology, mixture designs, sensory/instrumental correlation, and multivariate analysis.

Mr. Carr received his B.A. degree in Mathematics from the University of Dayton, and his Master's degree in Statistics from Colorado State University.

TABLE OF CONTENTS

Volume I

Volume II

Chapter 8

DESCRIPTIVE ANALYSIS TECHNIQUES

I. DEFINITION

All descriptive analysis methods involve the detection (discrimination) and the description of both the qualitative and quantitative sensory aspects of a product by trained panels of from 5 to 100 judges (subjects). Smaller panels of five to ten subjects are used for the typical product on the grocery shelf while the larger panels are used for products of mass production such as beers and soft drinks, where small differences can be very important.

Panelists must be able to detect and describe the perceived qualitative sensory attributes of a sample. These qualitative aspects of a product combine to define the product and include all of the appearance, aroma, flavor, texture, or sound properties of a product which differentiate it from others. In addition, panelists must learn to differentiate and rate the quantitative or intensity aspects of a sample and to define to what degree each characteristic or qualitative note is present in that sample. Two products may contain the same qualitative descriptors, but they may differ markedly in the intensity of each, thus resulting in quite different and easily distinctive sensory profiles or pictures of each product. The two samples below have the same qualitative descriptors, but they differ substantially in the amount of each characteristic (quantitatively). The numbers used represent intensity ratings on a 15-cm line scale on which a zero means no detectable amount of the attribute, and a 15 cm means a very large amount.[1]

The two samples (385 and 408) below are commercially available potato chips.

Characteristic	385	408
Fried potato	7.5	4.8
Raw potato	1.1	3.7
Vegetable oil	3.6	1.1
Salty	6.2	13.5
Sweet	2.2	1.0

Although these two samples of chips have the same attribute descriptors, they differ markedly by virtue of the amount of each flavor note. Sample 385 has a distinct fried potato and salt character with underlying oil, sweet, and raw potato notes. Sample 408 is dominated by saltiness with the potato, oil, and sweet notes of lesser impact.

II. FIELD OF APPLICATION

Use descriptive tests to obtain detailed descriptions of the aroma, flavor, and oral texture of foods and beverages, skinfeel of personal care products, handfeel of fabrics and paper products, and the appearance and sound of any product. These sensory pictures are used in research and development, and in manufacturing, to:

- Define the sensory properties of a target product for new product development[2]
- Define the characteristics/specifications for a control or standard for QA/QC and R&D applications
- Document a product's attributes before a consumer test to help in the selection of attributes to be included in the consumer questionnaire and to help in explanation of the results of the consumer test
- Track a product's sensory changes over time with respect to understanding shelflife, packaging, etc.

•   Map a product's perceived attributes for the purpose of relating them to instrumental, chemical, or physical properties

III. COMPONENTS OF DESCRIPTIVE ANALYSIS

A. Characteristics — The Qualitative Aspect

Those perceived sensory parameters which define the product are referred to by various terms such as attributes, characteristics, character notes, descriptive terms, descriptors, or terminology.

These qualitative factors (which are the same as the parameters discussed under classification, Chapter 5 Section II) include terms which define the sensory profile, picture, spectrum, or thumbprint of the sample. The selection of sensory attributes and the corresponding definition of these attributes should be related to the real chemical and physical properties of a product which can be perceived. Adherence to an understanding of the actual rheology or chemistry of a product makes the descriptive data easier to interpret and more useful for decision making.

The components of a number of different descriptive spectra are given below (a number of examples of each are shown in parentheses). Note that this list is also the key to a more complete list of descriptive terms given in Appendix 1 at the end of this chapter. The repeat appearance of certain properties and examples is intentional.

1. Appearance characteristics
    a. Color (hue, chroma, uniformity, depth)
    b. Surface texture (shine, smoothness/roughness)
    c. Size and shape (dimensions and geometry)
    d. Interactions among pieces or particles (stickiness, agglomeration, loose particles)
2. Aroma characteristics
    a. Olfactory sensations (vanilla, fruity, floral, skunky)
    b. Nasal feeling factors (cool, pungent)
3. Flavor characteristics
    a. Olfactory sensations (vanilla, fruity, floral, chocolate, skunky, rancid)
    b. Taste sensations (salty, sweet, sour, bitter)
    c. Oral feeling factors (heat, cool, burn, astringent, metallic)
4. Oral texture characteristics[4,5,6]
    a. Mechanical parameters, reaction of the product to stress (hardness, viscosity, deformation/fractureability)
    b. Geometrical properties, i.e., size, shape, and orientation of particles in the product (gritty, grainy, flaky, stringy)
    c. Fat/moisture properties, i.e., presence, release, and adsorption of fat, oil, or water (oily, greasy, juicy, moist, wet)
5. Skinfeel characteristics[7]
    a. Mechanical parameters, reaction of the product to stress (thickness, ease to spread, slipperiness, denseness)
    b. Geometrical parameters, i.e., size, shape, and orientation of particles in product or on skin after use (gritty, foamy, flaky)
    c. Fat/moisture parameters, i.e., presence, and absorption of fat, oil, or water (greasy, oily, dry, wet)
    d. Appearance parameters, visual changes during product use (gloss, whitening, peaking)
6. Texture/handfeel of woven and nonwoven fabrics
    a. Mechanical properties, reaction to stress (stiffness, force to compress or stretch, resilience)

    b. Geometrical properties, i.e., size, shape, and orientation of particles (gritty, bumpy, grainy, ribbed, fuzzy)

    c. Moisture properties, presence and absorption of moisture (dry, wet, oily, absorbent)

Again, the keys to the validity and reliability of descriptive analysis testing are

- Terms based on a thorough understanding of the technical and physiological principles of flavor or texture or appearance
- Thorough training of all panelists to fully understand the terms in the same way and to apply them in the same way (see Chapter 10)

B. Intensity — The Quantitative Aspect

The intensity or quantitative aspect of a descriptive analysis expresses the degree to which each of the characteristics (terms, qualitative components) is present. This is expressed by the assignment of some *value* along a measurement scale.

As with the validity and reliability of terminology, the validity and reliability of intensity measurements are highly dependent upon:

- The selection of a scaling technique which is broad enough to encompass the full range of parameter intensities and which has enough discrete points to pick up all the small differences in intensity between samples
- The thorough training of the panelists to use the scale in a similar way across all samples and across time (see Chapter 10 on panelist training)

Three types of scales are in common use in descriptive analysis:

1. *Category scales* are limited sets of words or numbers, constructed (as best one can) to maintain equal intervals between categories. A full description can be found in Chapter 5. A category scale from 0 to 9 is perhaps the most used in descriptive analysis, but longer scales are often justified. A good rule of thumb is to evaluate how many steps a panelist can meaningfully employ, and then to adopt a scale twice that long. Sometimes a 100-point scale is justified, e.g., in visual and auditory studies.

2. *Line scales* utilize a line 6 in. or 15 cm long on which the panelist makes a mark; they are described in Chapter 5. Line scales are almost as popular as category scales. Their advantage is that the intensity can be more accurately graded as there are no steps or "favorite numbers"; the chief disadvantage is that it is harder for a panelist to be consistent because a position on a line is not as easily remembered as a number.

3. *Magnitude estimation (ME) scales* are based on assignment of a number for the first sample tested, after which all subsequent numbers for subsequent samples are assigned in proportion, see Chapter 5. ME is used mostly in academic studies in which the focus is on a single attribute that can vary over a wide range of sensory intensities.[8,9]

Appendix 2 contains sets of reference samples useful for the establishment of scales for various odors and tastes, and also for the mechanical, geometrical, and moisture properties of oral texture. All the scales in Appendix 2 are based on a 15-cm line scale; however, the same standards can be distributed along a line or scale or any length or numerical value. The scales employ standard aqueous solutions such as sucrose, sodium chloride, citric acid, and caffeine, as well as certain widely available supermarket items which have shown adequate consistency, e.g., Hellman's Mayonnaise and Welch's Grape Juice.

C. Order of Appearance — The Time Aspect[10]

In addition to accounting for the attributes (qualitative) of a sample, and the intensity of each attribute (quantitative), panels can often detect differences among products in the order in which certain parameters manifest themselves. The order of appearance of physcial properties, related to oral, skin, and fabric textures, is generally predetermined by the way the product is handled (the input of forces by the panelist). By controlling the manipulation (one chew, one manual squeeze) the subject induces the manifestation of only a limited number of attributes (hardness, denseness, deformation) at a time.[11]

However, with the chemical senses (aroma and flavor), the chemical composition of the sample and some of its physical properties (temperature, volume, concentration) may alter the order in which certain attributes are detected.[12] Often in some products, such as beverages, the order of appearance of the characteristics is as indicative of the product profile as the individual aroma and flavor notes and their respective intensities.

Included as part of the treatment of the order of appearance of attributes is aftertaste or afterfeel, which includes those attributes that can still be perceived after the product or sample has been used or consumed. A complete picture of a product requires that all characteristics which are perceived after the product's use should be individually mentioned and rated for intensity.

Attributes described and rated for aftertaste or afterfeel do not necessarily imply a defect or negative note. For example, the cool aftertaste of a mouthwash or breath mint is a necessary and desirable property. On the other hand, a metallic aftertaste of a cola beverage may indicate a packaging contamination or a problem with a particular sweetener.

D. Overall Impression — The Integrated Aspect

In addition to the detection and description of the qualitative, quantitative, and time factors which define the sensory characteristics of a product, panelists are capable of and management is often interested in some integrated assessment of the product's properties. Ways in which such integration has been attempted include the following four.

Total intensity of aroma or flavor — A measure of the overall impact (intensity) of all the aroma components (perceived volatiles) or a measure of the overall flavor impact, which includes the aromatics, tastes, and feeling factors contributing to the flavor. Such an evaluation can be important in determining the general fragrance or flavor impact which a product delivers to the consumer, who does not normally understand all of the nuances of the contributing odors or tastes which the panel describes. The components of texture are more functionally discrete and "total texture" is not a property which can be determined.

Balance/blend (amplitude) — A well-trained descriptive panel is often asked to assess the degree to which various flavor or aroma characteristics fit together in the product. Such an evaluation involves a sophisticated understanding, half learned and half intuitive, of the appropriateness of the various attributes, their relative intensity in the complex, and the way(s) in which they harmonize in the complex. Evaluation of balance or blend (or amplitude, as it is called in the Flavor Profile method)[12,13] is difficult even for the highly trained panelist and should not be attempted with naive or less sophisticated subjects. In addition, care must be taken in the use of data on balance or blend. Often a product is not intended to be blended or balanced: a preponderance of spicy aromatics or toasted notes may be essential to the full character of a product. In some products the consumer may not appreciate a balanced composition, despite its well proportioned notes, as determined by the trained panel. Therefore it is important to understand the relative importance of blend or balance among consumers for the product in question before measuring and/or using such data.

Overall difference — In certain product situations the key decisions involve determination of the relative differences between samples and some control or standard product. Although the statistical analysis of differences between products on individual attributes is possible with many descriptive techniques, project leaders are often concerned with just how different a sample or prototype is from the standard. The determination of an overall difference (see Difference from Control Test, Chapter 6, Section II.F) allows the project management to make decisions regarding disposition of a sample based on its relative distance from the control; the accompanying descriptive information provides insight into the source and size of the relative attributes of the control and the sample.

Hedonic ratings — It is a temptation to ask the descriptive panel, once the description has been completed, to rate the overall acceptance of the product. In most cases this is a tempatation to be resisted, as the panel, through its training process, has been removed from the world of consumers and is no longer representative of any section of the general public. Some exceptions to this rule are listed in the introduction to Chapter 9. In the general case, training tends to change the personal preferences of panelists. As they become more aware of the various attributes of a product, panelists tend to assign weights to each attribute which are different from those of a regular consumer and which moreover can vary considerably from one panelist to the next, causing them to assign different "loadings" to different attributes.

IV. COMMONLY USED DESCRIPTIVE TEST METHODS

Over the last 40 years many descriptive analysis methods have been developed, and some have gained and maintained popularity as standard methods. The fact that these methods are described below is a reflection of their popularity, but it does not constitute a recommendation for use: on the contrary, a sensory analyst who needs to develop a descriptive system for a specific product and project application should study the literature on descriptive methods and should review several methods and combinations of methods before selecting the descriptive analysis system which can provide the most comprehensive, accurate, and reproducible description of each product and the best discrimination between products. See Section V below and review the I.F.T's Sensory Evaluation Guide,[12] which contains 109 references from different fields.

A. The Flavor Profile Method[13,14]

The Flavor Profile method was developed by Arthur D. Little, Inc. in the late 1940s. It involves the analysis of a product's perceived aroma and flavor characteristics, their intensities, order of appearance, and aftertaste, by a panel of four to six trained judges. An amplitude rating (see Section III above) is generally included as part of the profile.

Panelists are selected on the basis of a physiological test for taste discrimination, taste intensity discrimination, and olfactory discrimination and description. A personal interview is conducted to determine interest, availability, and potential for working in a group situation.

For training, panelists are provided with a broad selection of reference samples representing the product range, as well as examples of ingredient and processing variables for the product type. Panelists, with the panel leader's help in providing and maintaining reference samples, develop and define the terminology to be used in common by the entire panel. The panel also develops a common frame of reference for the use of the seven-point Flavor Profile intensity scale shown in Chapter 5.

The panelists, seated at a round or hexagonal table individually evaluate one sample at a time for both aroma and flavor and record the attributes (called "character notes"), their intensities, order of appearance, and aftertaste. Additional samples can be subsequently evaluated in the same session, but samples are not tasted back and forth. The results are reported to the panel leader, who then leads a general discussion of the panel to arrive at a "consensus" profile for each sample. The data are generally reported in tabular form, although a graphic representation is possible.

The Flavor Profile method may be applied when a panel must evaluate many different products, none of which is a major line of a major producer. The main advantage but also a major limitation of the Flavor Profile method is that it only used five to eight panelists. The lack of consistency and reproducibility which this entails is overcome to some extent by extra training and by the consensus method. However, the latter has been criticized for one-sidedness. The panel's opinion may become dominated by that of a senior member or a dominant personality, and equal input from other panel members is not obtained. Other points of criticism of the Flavor Profile are that screening methods do not include tests for the ability to discriminate specific aroma or flavor differences which may be of importance in specific product applications, and also that the seven-point scale limits the degree of discrimination between products showing small but important differences.

B. The Texture Profile Method[4-6]

Based somewhat on the principles of the Flavor Profile method, the Texture Profile method was developed by Brandt and Szczesniak at General Foods Corporation to define the textural parameters of foods. Later the method was expanded by Civille and Szczesniak[15] and Civille and Liska[11] to include specific attribute descriptors for specific products including semisolid foods, beverages,[24] skinfeel products (Schwartz,[7] Civille[24]), and fabric and paper goods.[24] In all cases the terminology is specific for each product type, but it is based on the underlying rheological properties expressed in the first Texture Profile publications.[4-6]

Panelists are selected on the basis of ability to discriminate known textural differences in the specific product application for which the panel is to be trained (solid foods, beverages, semisolids, skin care products, fabrics, paper, etc.). As with most other descriptive analysis techniques, panelists are interviewed to determine interest, availability, and attitude. Panelists, selected for training, are exposed to a wide range of products from the category under investigation, to provide a wide frame of reference. In addition, panelists are introduced to the underlying textural principles involved in the structure of the products under study. This learning experience provides panelists with understanding of the concepts of input forces and resulting strain on the product. In turn, panelists are able to avoid lengthy discussions about redundant terms and to select the most technically appropriate and descriptive terms for the evaluation of products. Panelists also define all terms and all procedures for evaluation, thus reducing some of the variability encountered in most descriptive testing. The reference scales used in the training of panelists can later serve as references for approximate scale values, thus further reducing panel variability.

Samples are evaluated independently by each panelist using one of the scaling techniques previously discussed. The original Texture Profile method used an expanded 13-point version of the Flavor Profile scale. In the last several years, however, Texture Profile panels have been trained using category, line, and ME scales (see Appendix 2 for food texture references for use with a 15-cm line scale). Depending on the type of scale used by the panel and on the way the data are to be treated, the panel verdicts may be derived by group consensus, as with the Flavor Profile method, or by statistical analysis of the data. For final reports the data may be displayed in tabular or graphic form.

C. The Quantitative Descriptive Analysis (QDA) Method[16,17]

In response to dissatisfaction among sensory analysts with the lack of statistical treatment of data obtained with the Flavor Profile or related methods, the Tragon Corporation developed the QDA method of descriptive analysis. This method relies heavily on statistical analysis to determine the appropriate terms, procedures, and panelists to be used for analysis of a specific product.

Panelists are selected from a large pool of candidates according to their ability to discriminate differences in sensory properties among samples of the specific product type for which they are to be trained. The training of QDA panels requires the use of product and ingredient references, as with other descriptive methods, to stimulate the generation of terminology. The panel leader acts as a facilitator, rather than as an instructor, and refrains from influencing the group. Attention is given to development of consistent terminology, but panelists are free to develop their own approach to scoring, using a 6-in. line scale which the method provides.

QDA panelists evaluate products one at a time in separate booths to reduce distraction and panelist interaction. The scoresheets are collected individually from the panelists as they are completed, and the data are entered for computation usually with a digitizer or card reader directly from the scoresheets. Panelists do not discuss data, terminology, or samples after each taste session and must depend on the discretion of the panel leader for any information on their performance relative to other members of the panel and to any known differences between samples.

The results of a QDA test are analyzed statistically, and the report generally contains a graphic representation of the data in the form of a "spider web" with a branch or spoke for each attribute.

The QDA method was developed in partial collaboration with the Department of Food Science at the University of California at Davis. It represents a large step towards the ideal of this book, the intelligent use of human subjects as measuring instruments, as discussed in Chapter 1. In particular, the use of a graphic scale, which reduces that part of the bias in scaling which results from the use of numbers, the statistical treatment of the data, the separation of panelists during evaluation, and the graphic approach to presentation of data, have done much to change the way in which sensory scientists and their clients view descriptive methodology. The following are areas which in our view could benefit from a change or further development:

1. The panel, because of lack of leadership, may develop erroneous terms. For example, the difference between natural vanilla and pure vanillin should be easily detected and described by a well-trained panel; however, an unguided panel would choose the term "vanilla" to describe the flavor of vanillin. Lack of direction also may allow a senior panelist or stronger personality to dominate the proceedings in all or part of the panel population.
2. The "free" approach to scaling can lead to inconsistency of results, partly because of particular panelists evaluating a product on a given day and not on another, and partly because of the context effects of one product seen after the other, with no external scale references.
3. The lack of immediate feedback to panelists on a regular basis reduces the opportunity for learning and expansion of terminology for greater capacity to discriminate and describe differences.
4. On a minor point, the practice of connecting "spokes" of the "spider web" can be misleading to some users, who by virtue of their technical training expect the area under a curve to have some meaning. In reality, the sensory dimensions shown in the "web" may be either unrelated to each other, or related in ways which cannot be represented in this manner.

D. Time-Intensity Descriptive Analysis[10]

In certain products a primary aspect of the descriptive process is the effect of time on the display or release of various characteristics. The order in which different sensory properties are manifested often defines the profile or thumbprint of the product as much as the characteristics and their intensity. The time of initial release of the mintiness of a chewing gum, as well as the length of time of flavor release, have significant effects on consumer preference.

The order in which one spaghetti sauce reveals its herbs and spices clearly differentiates it from another.

Time-intensity descriptive analysis involves the monitoring of specific attributes and their intensities over time. In some cases the panelists are asked to rate the intensities of 3 to 5 attributes every 10 or 15 sec following consumption or use. Products such as confections, skincare lotions and creams, and fabrics, which change with handling over time, require a time-intensity system for full description. When one attribute, such as the bitterness of beer[18-20] or the sweetness of an artificial sweetener[21] requires critical monitoring over time, a stripchart recorder or a microcomputer[22] can be used to collect the data as a continuous sequence.

V. DESIGNING A DESCRIPTIVE PROCEDURE. THE SPECTRUM™ METHOD*

The name Spectrum™ covers a procedure, designed by one of us (G.V.C.) and developed over several years of collaboration with a number of large companies. The basic philosophy of Spectrum™ is pragmatic: it provides the tools with which to design a descriptive procedure for a given product. The principal tools are the reference lists contained in Appendixes 1 and 2, together with the scaling procedures and methods of panel training described in Chapters 5 and 10. The aim is to choose the most practical system given the product in question, the overall sensory program, the specific project objective(s) in developing a panel, and the desired level of statistical treatment of the data.

For example, panelists may be selected and trained to evaluate only one product, or a variety of products. Products may be described in terms of only appearance, aroma, flavor, or texture characteristics, or panelists may be trained to evaluate all of these attributes. In short, Spectrum™ is a "custom design" approach to panel development, selection, training, and maintenance.

Courses teaching the basic elements of Spectrum are available and a detailed manual is in preparation.**

A. Terminology

Terms may be chosen from Appendix 1 using the key in Section III.A. The choice of terms may be broad or narrow according to the panel's objective — only aroma characteristics, or all sensory modalities. However, the method requires that all terminology is developed and derived by a panel which has been exposed to the underlying technical principles of each modality to be described. For example, a panel describing color must understand color intensity, color hue, and chroma. A panel involved in oral, skinfeel, and/ or fabric texture needs to understand the effects of rheology and mechanical characteristics and how these in turn are affected by moisture level and particle size. The chemical senses pose an even greater challenge in requiring panelists to demonstrate a valid response to changes in ingredients and processing. Words such as vanilla, chocolate, and orange must describe an authentic vanilla, chocolate, and orange character for which concrete references are supplied. Impressions of vanillin, cocoa, and distilled orange oil require separate terms and references. If the panel hopes to attain the status of "expert panel" in a given field, it must demonstrate that it can use a concrete list of descriptors based on an understanding of the underlying technical differences among samples of a product.

---

*   Application for Trade Name in progress.
** G. V. Civille, Sensory Spectrum, Inc., 44 Brentwood Drive, East Hanover, NJ 07936; Center for Professional Advancement, P.O. Box H., East Brunswick, NJ 08816-0257.

B. Intensity

Different project objectives may require widely different intensity scales. The key property is the number of points of discrimination along the scale. If product differences require a large number of points of discrimination to clearly define intensity differences both within and between attributes, the panel leader requires a 15-cm line scale, or a category scale with 30 points or more, or a ME scale.

The Spectrum™ method is based on extensive use of reference points, which may be chosen according to the guidelines given in Appendix 2. These are derived from the collective data of several panels over several replicates. Whatever the scale chosen, it must have at least two and preferably three to five reference points distributed across the range. A set of well-chosen reference points greatly reduces panel variability, allowing for comparison of data across time and products. Such data also allow more precise correlation with stimulus changes (stimulus/response curves) and with instrumental data (sensory/instrumental correlations). The choice of a scaling technique may also depend on the available facilities for computer manipulation of data and on the need for sophisticated data analysis.

C. Other Options

Included among the tools of the Spectrum™ method are time/intensity tests, difference from control tests, total flavor impact assessment, and others. The basic philosophy, as mentioned, is to train the panel to fully define each and all of a product's sensory attributes, to rate the intensity of each, and to include other relevant characterizing aspects such as changes over time, differences in the order of appearance of attributes, and integrated total aroma and/or flavor impact.

The creative and diligent sensory analyst can construct the optimal descriptive technique by selecting from the spectrum of terms, scaling techniques and other optional components which are available at the start of each panel development.

VI. MODIFIED SHORT VERSION DESCRIPTIVE SPECTRUM™ PROCEDURES FOR QUALITY ASSURANCE, SHELF-LIFE STUDIES, ETC.

Certain applications of descriptive analysis require evaluation of a few detailed attributes without a full analysis of all the parameters of flavor, texture, and/or appearance. The tracking or monitoring of product changes, necessary in QC/QA sensory work and in shelf-life studies, can provide the required information from data on a small number of selected sensory properties over time. The modified or short-version descriptive technique, in any situation, must be based on work done with a fully trained descriptive panel, generally in R&D, which characterizes all of the product's attributes. Once the panel has evaluated a succession of products typical of the full range of sensory properties, e.g., several production samples from all plants and through the practical ages and storage conditions encountered, the sensory analyst and project team can determine five to ten key parameters, which define the span of qualities from "typical" to "off". Monitoring of these attributes permits QA/QC and R&D to identify any changes that require troubleshooting and correction.

Maximum benefit can be derived from this system when it is coupled with a Difference-from-Control test. The modified descriptive panel is trained to recognize the control or standard product along with other samples which the fully trained panel has described as different from the control on the key attributes. The panel is shown the full range of samples and asked to rate them using the normal Difference-from-Control scale (see Chapter 6, Section II.F). The panel understands that occasionally one of the test samples during normal testing of production will be a blind control and/or one of the original "small difference" or "large difference" demonstration samples. This precaution reduces the likelihood of panelists anticipating too much change in shelf-life studies, or too little change in production.

The Difference from Control test provides an indication of the magnitude of the difference from the standard product. Samples may on occasion show statistical significance for a difference from the control and yet remain acceptable to consumers. The product team can submit to consumer testing three or more products, identified by the panel as showing slight, moderate, and large differences from the control. In place of a ''go''/''no go'' system based strictly on statistical significance, the company can devise a system of specifications based on known differences that are meaningful to the consumer. The system can be used to track production or storage samples over time in a cost effective program (see Chapter 9, Section VI, Example 5).

APPENDIX 1

SPECTRUM® REFERENCE LISTS OF TERMINOLOGY FOR DESCRIPTIVE ANALYSIS

The following lists of terms for appearance, flavor, and texture can be used by panels suitably trained, to define the qualitative aspects of a sample.

When required, each of the terms can be quantified using a scale chosen from Chapter 5. Each scale must have at least two and preferably three to five reference points chosen, e.g., from Appendix 2.

A simple scale can have general anchors:

None---Strong

or a scale can be anchored using bipolar words (opposites):

Smooth---Lumpy
Soft--Hard

Attributes perceived via the chemical senses in general use a unipolar intensity scale (None-Strong), while for appearance and texture attributes, a bipolar scale is best, as shown below.

A. Terms Used to Describe Appearance

1. Color
    a. Description   The actual color name or hue, such as red, blue, etc. The description can be expressed in the form of a scale range, if the product covers more than one hue:
    [Red --- Orange]
    b. Intensity   The intensity or strength of the color from light to dark:
    [Light --- Dark]
    c. Brightness   The chroma (or purity) of the color, ranging from dull, muddied to pure, bright color. Fire engine red is a brighter color than burgundy red.
    [Dull --- Bright]
    d. Evenness   The evenness of distribution of the color, not blotchy:
    [Uneven/blotchy --- Even]

2. Consistency/
   Texture

a.	Thickness	The viscosity of the product: [Thin --- Thick]
b.	Roughness	The amount of irregularity, protrusions, grains, or bumps which can be seen on the surface of the product; smoothness is the absence of surface particles: [Smooth --------------------------- -------------------------- Rough] Graininess is caused by small surface particles: [Smooth -- Grainy] Bumpiness is caused by large particles: [Smooth --- Bumpy]
c.	Particle interaction (Stickiness): (Clumpiness):	The amount of stickiness among particles or the amount of agglomeration of small particles: [Not sticky --- Sticky] [Loose particles --- Clumps]

3. Size/shape

a.	Size	The relative size of the pieces or particles in the sample: [Small --- Large] [Thin --- Thick]
b.	Shape	Description of the predominant shape of particles: flat, round, spherical, square, etc. [No scale]
c.	Even distribution	Degree of uniformity of particles within the whole: [Nonuniform pieces ----------------------------- Uniform pieces]

4. Surface shine

Amount of light reflected from the product's surface:
[Dull ------------- -- Shiny]

B. General Flavor Terms

The full list of fragrance and flavor descriptors is too unwieldy to reproduce here; the list of aromatics* alone contains over a thousand words. In the following, aromatics for Baked Goods are shown as an example.

Flavor is the combined effects of the

- Aromatics
- Tastes
- Chemical feelings

stimulated by a substance in the mouth. For Baked Goods it is convenient to subdivide the aromatics into

- Grainy aromatics
- Grain-related terms
- Dairy terms
- Other processing characteristics
- Sweet aromatics
- Added flavors/aromatics
- Aromatics from shortening
- Other aromatics

* The term Aromatics is used in this book to cover that portion of the flavor which is perceived by the sense of smell, from a substance in the mouth.

1.   Aromatics (of
       baked goods)
     a.   Grainy aromatics   Those aromatics or volatiles which are derived from various grains; the term Cereal can be used as an alternative, but it implies finished and/or toasted character and is, therefore, less useful than grainy.

Grainy: The general term to describe the aromatics of grains, which cannot be tied to a specific grain by name.

Terms pertaining to a specific grain: corn, wheat, oat, rice, soy, rye.

Grain character modified or characterized by a processing note, or lack thereof:

Raw corn	Cooked corn	Toasted corn
Raw wheat	Cooked wheat	Toasted wheat
Raw oat	Cooked oat	Toasted oat
Raw rice	Cooked rice	Toasted rice
Raw soy	Cooked soy	Toasted soy
Raw rye	Cooked rye	Toasted rye

Definitions of processed grain terms:

Raw (name) flour: The aromatics perceived in a particular grain which has not been heat treated.

Cooked (name) flour: The aromatics of a grain which has been gently heated or boiled; Cream of Wheat has cooked wheat flavor; oatmeal has cooked oat flavor.

Baked toasted (name) flour: The aromatics of a grain which has been sufficiently heated to caramelize some of the starches and sugars.

     b.   Grain-related
            terms
Green: The aromatic associated with unprocessed vegetation, such as fruits and grains; this term is related to Raw, but has the additional character of hexenals, leaves, and grass.

Hay-like/grassy: Grainy aromatic with some green character of freshly mowed, air-dried grain or vegetation.

Malty: The aromatics of toasted malt.

     c.   Dairy terms   Those volatiles related to milk, butter, cheese and other cultured dairy products. This group includes the following terms:

Dairy: As above.

Milky: More specific than Dairy, the flavor of regular or cooked cow's milk.

Buttery: The flavor of high-fat fresh cream or fresh butter; not rancid, butyric or diacetyl-like.

Cheesy: The flavor of milk products treated with rennet which hydrolyzes the fat, giving them a butyric or isovaleric acid character.

     d.   Other processing
            characteristics
Caramelized: A general term used to describe starches and sugars which have been browned; used alone when the starch or sugar (e.g., toasted corn) cannot be named.

Burnt: Related to overheating, overtoasting, or scorching the starches or sugars in a product.

     e.   Added flavors/
            aromatics
The following terms relate to specific ingredients which may be added to baked goods to impart specific character notes; in each case, references for the term are needed:

Nutty: peanut, almond, pecan, etc.

Chocolate: milk chocolate, cocoa, chocolate-like.

Spices: cinnamon, clove, nutmeg, etc.

Yeasty: natural yeast (not chemical leavening).

f. Aromatics from shortening   The aromatics associated with oil or fat based shortening agents used in baked goods:

Buttery: See Dairy above.

Oil flavor: The aromatics associated with vegetable oils, not to be confused with an oily film on the mouth surfaces, which is a texture characteristic.

Lard flavor: The aromatics associated with rendered pork fat.

Tallowy: The aromatics associated with rendered beef fat.

g. Other aromatics: The aromatics which are not usually part of the normal product profile and/or do not result from the normal ingredients or processing of the product:

Vitamin: Aromatics resulting from the addition of vitamins to the product.

Cardboard flavor: Aromatics associated with the odor of cardboard box packaging, which could be contributed by the packaging *or* by other sources, such as staling flours.

Rancid: Aromatics associated with oxidized oils, often also described as painty or fishy.

Mercaptan: Aromatics associated with the mercaptan class of sulfur compounds. Other terms which panelists may use to describe odors arising from sulfur compounds are skunky, sulfitic, rubbery.

End of section referring to Baked Goods only.

2. Basic tastes

a. Sweet   The taste stimulated by sucrose and other sugars, such as fructose, glucose, etc. and by other sweet substances such as saccharin, aspartame, and acesulfam K.

b. Sour   The taste stimulated by acids, such as citric, malic, phosphoric, etc.

c. Salty   The taste stimulated by sodium salts, such as sodium chloride and sodium glutamate, and in part by other salts, such as potassium chloride.

d. Bitter   The taste stimulated by substances such as quinine, caffeine, and hop bitters.

3. Chemical feeling factors   Those characteristics which are the response of tactile nerves to chemical stimuli.

a. Astringency   The shrinking or puckering of the tongue surface caused by substances such as tannins or alum.

b. Heat   The burning sensation in the mouth caused by certain substances such as capsaicin from red or piperin from black peppers; mild heat or warmth is caused by some brown spices.

c. Cooling   The cool sensation in the mouth or nose produced by substances such as menthol and mints.

C. Terms Used to Describe Oral Texture (With Procedures and Definitions)
Each set of texture terms includes the procedure for manipulation of the sample.

1.	Surface texture	Feel surface of sample with lips and tongue.
		The overall amount of small and large particles in the surface:
	a. Geometrical in surface	[Smooth --- Rough]
		Large particles: amount of bumps/lumps in surface:
		[Smooth --- Bumpy]
		Small particles: amount of small grains in surface:
		[Smooth --- Grainy]
		Crumbly: amount of loose, grainy particles free of the surface:
	b. Loose geometrical	[None --- Many]
		The amount of wetness or oiliness (moistness if both) on surface:
	c. Moistness/ dryness	[Dry --- Wet/oily/moist]
2.	Partial compression	Compress partially (specify with tongue, incisors, or molars) without breaking, and release.
	a. Springiness (rubberiness)	Degree to which sample returns to original shape after a certain time period:
		[No recovery ------------------------------------- Very springy]
3.	First bite	Bite through a predetermined size sample with incisors.
	a. Hardness	Force required to bite through:
		[Very soft --- Very hard]
	b. Cohesiveness	Amount of sample that deforms rather than ruptures:
		[Breaks --- Deforms]
	c. Fracturability	The force with which the sample breaks:
		[Crumbles --- Fractures]
	d. Uniformity of bite	Evenness of force throughout bite:
		[Uneven, choppy ------------------------------------- Very even]
	e. Moisture release	Amount of wetness/juiciness released from sample:
		[None -- Very juicy]
	f. Geometrical	Amount of particles resulting from bite, or detected in center of sample:
		[None -------------------------- Very grainy (gritty, flaky, etc.)]
4.	First chew	Bite through a predetermined size sample with molars.
	a. Hardness	As above:
		[Very soft --- Very Hard]
	b. Cohesiveness/ fracturability	Both as above:
		[Breaks --- Deforms]
		[Crumbles --- Fractures]
	c. Adhesiveness	Force required to remove sample from molars:
		[Not sticky --- Very sticky]
	d. Denseness	Compactness of cross section:
		[Light/airy -- Dense]
	e. Geometrical	As above:
		[None -------------------------- Very grainy (gritty, flaky, etc.)]
	f. Moist/moisture release	As above:
		[None -- Very juicy]

5. Chew down
    Chew sample with molars for a predetermined number of chews (enough to mix sample with saliva to form a mass).

    a. Moisture absorption
    Amount of saliva absorbed by product:
    [None -- All]

    b. Cohesiveness of mass
    Degree to which sample holds together in a mass:
    [Loose mass -------------------------------------- Compact mass]

    c. Adhesiveness of mass
    Degree to which mass sticks to the roof of the mouth or teeth:
    [Not sticky --------------------------------------- Very sticky]

6. Rate of melt
    (When applicable): amount of product melted after a certain number of chews:
    [None -- All]

    a. Geometrical in mass
    Roughness/graininess/lumpiness: Amount of particles in mass:
    [None --- Many]

    b. Moistness of mass
    Amount of wetness/oiliness/moisture in mass:
    [Dry -- Moist/Oily/Wet]

    c. Number of chews to disintegrate
    Count number.

7. Residual
    Swallow sample.

    a. Geometrical
    (Chalky, particles): amount of particles left in mouth:
    [None -- Very much]

    b. Oily mouth coating
    Amount of oil left on mouth surfaces:
    [None -- Very much]

    c. Sticky mouth coating
    Stickiness/tackiness of coating when tapping tongue on roof of mouth:
    [Not sticky --- Very sticky]

    d. Tooth packing
    Amount of product left on teeth:
    [None -- Very much]

D. Example of Texture Terminology: Oral Texture of Cookies

Surface:
Place cookie between lips and evaluate for:
Roughness: Degree to which surface is uneven [Smooth---Rough]
Loose particles: Amount of loose particles on surface
[None --- Many]
Dryness: Absence of oil on the surface [Oily --- Dry].

First bite:
Place one third of cookie between incisors, bite down, and evaluate for:
Fracturability: Force with which sample ruptures:
[Crumbly --- Brittle]
Hardness: Force required to bite through sample [Soft --- Hard]
Particle size: Size of crumb pieces: [Small --- Large]

First chew:
Place one third of cookie between molars, bite through, and evaluate for:
Denseness: Compactness of cross section: [Airy --- Dense]
Uniformity of Chew: Degree to which chew is even throughout:
[Uneven --- Even]

Chew down:
Place one third of cookie between molars, chew 10 to 12 times, and evaluate for:
Moisture absorption: Amount of saliva absorbed by sample:
[None --- Much]

Type of breakdown: Thermal, mechanical, salivary
[No scale]
Cohesiveness of mass: Degree to which mass holds together:
[Loose --- Cohesive]
Tooth Pack: Amount of sample stuck in molars: [None --- Much]
Grittiness: Amount of small, hard particles between teeth during chew:
[None --- Many]

Residual: Swallow sample and evaluate residue in mouth:
Oily: Degree to which mouth feels oily: [Dry --- Oily]
Particles: Amount of particles left in mouth: [None --- Many]
Chalky: Degree to which mouth feels chalky:
[Not chalky --- Very chalky]

APPENDIX 2

REFERENCE SAMPLES USEFUL FOR THE ESTABLISHMENT OF SPECTRUM™ INTENSITY SCALES FOR DESCRIPTIVE ANALYSIS

The scales below (all run from 0 to 15) contain intensity values for aromatics (A), and for tastes and chemical feeling factors (B), which were derived from repeated tests with trained panels at Hill Top Research, Inc., Cincinnati, Ohio, and also for various texture characteristics (C), which were obtained from repeated tests at Hill Top Research or were developed at Best Foods Division, CPC International, Union, N.J.[23]

New panels can be oriented to the use of the 0 to 15 scale by presentation of the concentrations of caffeine, citric acid, NaCl and sucrose which are listed under Section B. If a panel is developing a descriptive system for an orange drink product, the panel leader can present three "orange" references:

1. Fresh squeezed orange juice labeled "Orange Complex 7.5"
2. Reconstituted Minute Maid concentrate labeled "Orange Complex 6.5 *and* Orange Peel 3.0"
3. Tang labeled "Orange Complex 9.5 *and* Orange Peel 9.5"

At each taste test of any given product, labeled reference samples related to its aromatic complex can be presented, so as to standardize the panel's scores and keep them from drifting.

A. Intensity Scale Values (0 to 15) for Some Common Aromatics

Term	Reference	Scale value
Astringency	Grape juice (Welch)	6.5
	Tea bags/1 hr soak	6.5
Baked wheat	Sugar Cookies (Kroger)	4
	Brown Edge Cookies (Nabisco)	5
Baked white wheat	Ritz Crackers (Nabisco)	6.5
Caramelized sugar	Brown Edge Cookies (Nabisco)	3
	Sugar Cookies (Kroger)	4
	Social Tea (Nabisco)	4
	Bordeaux Cookies (Pepperidge Farm)	7

A. Intensity Scale Values (0 to 15) for Some Common Aromatics (continued)

Term	Reference	Scale value
Celery	V-8 (Campbell)	5
Cheese	American Cheese (Kraft)	5
Cinnamon	Big Red Gum (Wrigley)	12
Cooked apple	Apple Sauce (Mott)	5
Cooked milk	Butterscotch Pudding (Royal)	4
Cooked orange	Frozen Orange Concentrate (Minute Maid) — reconstituted	5
Cooked wheat	Pasta (DeCecco) — cooked	5.5
Egg	Mayonnaise (Hellmann's)	5
Egg flavor	Hard boiled egg	13.5
Grain complex	Cream of Wheat (Nabisco)	4.5
	Spaghetti (DeCecco) — cooked	4.5
	Ritz Cracker (Nabisco)	6
	Whole wheat spaghetti (DeCecco) — cooked	6.5
	Triscuit (Nabisco)	8
	Wheatina Cereal	9
Grape	Kool-Aid	4.5
	Grape juice (Welches)	10
Grapefruit	Bottled grapefruit juice (Kraft)	10
Lemon	Brown Edge Cookies (Nabisco)	3
	Lemonade (Country Time)	5
Milky complex	American cheese (Kraft)	3
	Powdered milk (Carnation)	4
	Whole milk	5
Mint	Doublemint Gum (Wrigley)	11
Oil	Potato chips (Pringles)	1
	Potato chips (Frito Lay)	2
	Heated oil (Crisco)	5.5
Orange complex	Orange drink (Hi-C)	3
	Frozen orange concentrate (Minute Maid) — reconstituted	6.5
	Fresh squeezed orange juice	7.5
	Orange concentrate (Tang)	9.5
Orange peel	Soda (Orange Crush)	2
	Frozen orange concentrate (Minute Maid) — reconstituted	3
	Orange Concentrate (Tang)	9.5
Peanut, med. roasted	(Planters)	7
Potato	Potato chips (Pringles)	4.5
Roastedness	Coffee (Maxwell House)	7
	Espresso Coffee (Medaglia D'Oro)	14
Soda	Saltines (Nabisco)	2
Spice complex	Spice cake (Sara Lee)	7.5
Tuna	Canned light tuna (Bumble Bee)	11
Vanillin	Sugar cookies (Kroger)	7

B. Intensity Scale Values (0 to 15) for the Four Basic Tastes in Various Products

	Sweet	Salt	Sour	Bitter
American cheese (Kraft)		7	5	
Applesauce, natural (Motts)	5		4	
Applesauce, regular (Motts)	8.5		2.5	
Big Red Gum (Wrigley)	11.5			
Bordeaux Cookies (Pepperidge Farm)	12.5			
Caffeine, solution in water				
0.05%				2
0.08%				5
0.15%				10
0.20%				15
Celery seed				9
Chocolate bar (Hershey)	10		5	4
Citric acid, solution in water				
0.05%			2	
0.08%			5	
0.15%			10	
0.20			15	
Coca Cola Classic	9			
Endive, raw				7
Fruit punch (Hawaian)	10		3	
Grape juice (Welch)	6		7	2
Grape Kool-Aid	10		1	
Grapefruit juice, bottled (Kraft)	3.5		13	2
Kosher dill pickle (Vlasic)		12	10	
Lemon juice (Realemon)			15	
Lemonade (Country Time)	7		5.5	
Mayonnaise (Hellman's)		8	3	
NaCl, solution in water				
0.2%		2		
0.5%		5		
1.0%			10	
1.5%			15	
Orange (fresh squeezed juice)	6		7.5	
Soda (Orange Crush)	10.5		2	
Frozen orange concentrate (Minute Maid) — reconstituted	5.5		5	
Potato chips (Frito Lay)		9.5		
Potato chips (Pringles)		8.5		
Ritz cracker (Nabisco)	4	8		
Soda cracker (Premium)		5		
Spaghetti sauce (Ragu)	8	12		
Sucrose, solution in water				
2.0%	2			
5.0%	5			
10.0%	10			
16.0%	15			
Sweet pickle (Vlasic)	8.5		8	

B. Intensity Scale Values (0 to 15) for the Four Basic Tastes in Various Products (continued)

	Sweet	Salt	Sour	Bitter
Orange concentrate (Tang)	9.5		4.5	
Tea bags soaked 1 hr				8
Triscuit (Nabisco)		9.5		
V-8 vegetable juice (Campbell)	8			
Wheatina Cereal	6			2.5

C. Intensity Scale Values (0 to 15) for Some Common Texture Attributes

*1. Standard Roughness Scale**

Scale value	Reference	Brand/type/ manufacturer	Sample size
0.0	Gelatin dessert	Jello	2 tbsp.
5.0	Orange peel	Peel from fresh orange	$\frac{1}{2}$" piece
8.0	Potato chips	Pringles	5 pieces
12.0	Hard granola bar	Quaker Oats	$\frac{1}{2}$ bar
15.0	Rye wafer	Finn Crisp	$\frac{1}{2}$" square

Technique: Hold sample in mouth; feel the surface to be evaluated with the lips and tongue.
Definition: The amount of particles *in* the surface:
[Smooth --- Rough]

2. Standard Wetness Scale

Scale value	Reference	Brand/type/manufacturer	Sample size
0.0	Unsalted Premium cracker	Nabisco	1 cracker
3.0	Carrots	Uncooked, fresh, unpeeled	$\frac{1}{2}$" slice
7.5	Apples	Red Delicious, uncooked, fresh, unpeeled	$\frac{1}{2}$" slice
10.0	Ham	Oscar Mayer	$\frac{1}{2}$" piece
15.0	Water	Filtered, room temp.	$\frac{1}{2}$ tbsp.

Technique: Hold the sample in mouth; feel surface with lips and tongue.
Definition: The amount of moisture, due to an aqueous system, on the surface:
[Dry -- Wet]

3. Standard Stickiness to Lips Scale

Scale value	Reference	Brand/type/manufacturer	Sample size
0.0	Cherry tomato	Uncooked, fresh, unpeeled	$\frac{1}{2}$" slice
4.0	Nougat	Three Musketeers/M & M Mars	$\frac{1}{2}$" cake Remove chocolate first
7.5	Breadstick	Stella d'Oro	$\frac{1}{2}$ stick
10.0	Pretzel rod	Bachmans	1 piece
15.0	Rice Krispies	Kellogg's	1 tsp.

Technique: Hold sample near mouth; compress sample lightly between lips and release.
Definition: The degree to which the surface of the sample adheres to the lips:
[None --- Very]

---

* The roughness scale measures the amount of irregular particles *in* the surface. These may be small (chalky, powdery), medium (grainy), or large (bumpy).

4. Standard Springiness Scale

Scale value	Reference	Brand/type/manufacturer	Sample size
0.0	Cream cheese	Philadelphia/Kraft	$1/2$" cube
5.0	Frankfurter	Cooked 10 min/Hebrew National	$1/2$ slice
9.5	Marshmallow	Minature Marshmallow/Kraft	3 pieces
15.0	Gelatin dessert	Jello, Knox (Note 1)	$1/2$" cube

Technique: Place sample between molars; compress partially without breaking the sample structure; release.

Definition: (1) The degree to which sample returns to original shape or
(2) the rate with which sample returns to original shape:
[Not springy -- Very springy]

Note 1: One package Jello and one package Knox gelatin are dissolved in $1 1/2$ cups hot water and refrigerated for 24 hr.

5. Standard Hardness Scale

Scale value	Reference	Brand/type/manufacturer	Sample size
1.0	Cream cheese	Philadelphia/Kraft	$1/2$" cube
2.5	Egg white	Hard cooked	$1/2$" cube
4.5	Cheese	Yellow American pasteurized process/Land O'Lakes	$1/2$" cube
6.0	Olives	Goya Foods/giant size, stuffed	1 olive pimento removed
7.0	Frankfurter	Large, cooked 5 min/Hebrew National	$1/2$" slice
9.5	Peanuts	Cocktail type in vacuum tin/ Planters	1 nut, whole
11.0	Carrots	Uncooked, fresh, unpeeled	$1/2$" slice
11.0	Almonds	Shelled/Planters-Nabisco	1 nut
14.5	Hard candy	Life Savers	3 pieces, one color

Technique: For solids, place food between the molars and bite down evenly, evaluating *the force required to compress the food.* For semisolids, measure hardness by compressing the food against palate with tongue. When possible, the sample height for hardness standards is $1/2$".

Definition: The force to attain a given deformation, such as:
- force to compress between molars, as above
- force to compress between tongue and palate
- force to bite through with incisors

[Soft -- Hard]

6. Standard Cohesiveness Scale

Scale value	Reference	Brand/type/manufacturer	Sample size
1.0	Corn muffin	Pepperidge Farm	$1/2$" cube
5.0	Cheese	Yellow American pasteurized American/Land O'Lakes	$1/2$" cube
8.0	Pretzel	Soft pretzel	$1/2$ stick
10.0	Dried fruit	Sun Dried Seedless Raisins/ Sun-Maid	1 tsp.
12.5	Candy chews	Starburst/M & M-Mars	1 piece
15.0	Chewing gum	Freedent	1 stick

Technique: Place sample between molars; compress fully (can be done with incisors).

Definition: The degree to which sample deforms rather than crumbles, cracks, or breaks:
[Rupturing -- Deforming]

7. Standard Fracturability Scale

Scale value	Reference	Brand/type/manufacturer	Sample size
1.0	Corn muffin	Thomases	$^1/_2$" cube
2.5	Egg Jumbos	Stella D'Oro	$^1/_2$" cube
4.2	Graham crackers	Nabisco	$^1/_2$" square
6.7	Melba toast	Plain, rectangular/Devonsheer, Melba Co.	$^1/_2$" square
8.0	Ginger Snaps	Nabisco	$^1/_2$" square
10.0	Rye Wafers	Finn Crisp/Shaffer, Clark & Co.	$^1/_2$" square
13.0	Peanut brittle	Kraft	$^1/_2$" square candy part
14.5	Life Savers	Nabisco	1 piece

Technique: Place food between molars and bite down evenly until the food crumbles, cracks, or shatters.

Definition: The force with which the sample breaks:
[Crumbly -- Brittle]

8. Standard Viscosity Scale

Scale value	Reference	Brand/type/manufacturer	Sample size
1.0	Water	Bottled Mountain Spring	$^1/_2$ tsp.
2.2	Light cream	Sealtest Foods	$^1/_2$ tsp.
3.0	Heavy cream	Sealtest Foods	$^1/_2$ tsp.
3.9	Evaporated milk	Carnation Co.	$^1/_2$ tsp.
6.8	Maple syrup	Vermont Maid, R. J. Reynolds	$^1/_2$ tsp.
9.2	Chocolate syrup	Hershey Chocolate	$^1/_2$ tsp.
11.7	mixture: $^1/_2$ cup condensed milk + 1 T. heavy cream	Magnolia Sweetened Borden Foods	$^1/_2$ tsp.
14.0	Condensed milk	Borden Foods	$^1/_2$ tsp.

Technique: (A) Place 1 tsp. of product close to lips; draw air in gently to induce flow of liquid; measure the force required.
(B) Once product is in mouth, allow to flow across tongue by moving tongue slowly to roof of mouth; measure rate of flow (the force here is gravity).

Definition: The rate of flow per unit force:
(A) the force to draw between lips from spoon
(B) the rate of flow across tongue.
[Not viscous --- Viscous]

9. Standard Denseness Scale

Scale value	Reference	Brand/type/manufacturer	Sample size
0.5	Cool Whip	Birds Eye/General Foods	2 tbsp.
2.5	Marshmallow Fluff	Fluff-Durkee-Mower	2 tbsp.
4.0	Nougat	Three Musketeers/M&M Mars	$^1/_2$" cube Remove chocolate first
6.0	Malted milk balls	Whopper, Leaf Confectionery	5 pieces
9.0	Frankfurter	Cooked 5 min, Oscar Mayer	5 $^1/_2$" slices
13.0	Fruit jellies	Chuckles/Nabisco	3 pieces

Technique: Place sample between molars and compress.

Definition: The compactness of the cross section:
[Airy -- Dense]

10. Standard Moisture Absorption Scale

Scale value	Reference	Brand/type/manufacturer	Sample size
0.0	Licorice	Shoestring	1 piece
4.0	Licorice	Twizzlers/Red Licorice/Hershey	1 piece
7.5	Popcorn	Bagged popcorn/Bachman	2 tbsp.
10.0	Potato chips	Wise	2 tbsp.
13.0	Cake	Pound Cake, frozen type/Sara Lee	1 slice
15.0	Saltines	Unsalted top premium cracker/ Nabisco	1 cracker

Technique: Chew sample with molars for up to 15 to 20 chews.
Definition: The amount of saliva absorbed by sample during chew down:
              [No absorption -- Large amount of absorption]

11. Standard Cohesiveness of Mass Scale

Scale value	Reference	Brand/type/manufacturer	Sample size
0.0	Licorice	Shoestring	1 piece
2.0	Carrots	Uncooked, fresh, unpeeled	$^1/_2''$ slice
4.0	Mushroom	Uncooked, fresh	$^1/_2''$ slice
7.5	Frankfurter	Cooked 5 min/Hebrew National	$^1/_2''$ slice
9.0	Cheese	Yellow American pasterized process/Land O'Lakes	$^1/_2''$ cube
13.0	Soft brownie	Archway Cookies	$^1/_2''$ cube
15.0	Dough	Pillsbury Country/Biscuit Dough	1 tbsp.

Technique: Chew sample with molars for up to 15 chews.
Definition: The degree to which chewed sample (at 10 to 15 chews) holds together in a mass:
              [Loose mass -- Tight mass]

12. Standard Tooth Packing Scale

Scale value	Reference	Brand/type/manufacturer	Sample size
0.0	Mini-clams	Geisha/Nozaki America	3 pieces
1.0	Carrots	Uncooked, fresh, unpeeled	$^1/_2''$ slice
3.0	Mushrooms	Uncooked, fresh, unpeeled	$^1/_2''$ slice
7.5	Cracker	Graham cracker/Nabisco	$^1/_2''$ square
9.0	Cheese	Yellow American pasteurized process/Land O'Lakes	$^1/_2''$ cube
11.0	Cheese Snacks	Wise-Borden Cheese Doodles	5 pieces
15.0	Candy	Ju-Jubes	3 pieces

Technique: After sample is swallowed, feel the tooth surfaces with tongue.
Definition: The degree to which product sticks on the surface of teeth:
              [None stuck --- Very much stuck]

REFERENCES

1. **Civille, G. C.,** *Descriptive Analysis,* (course notes for sensory short course), Instutite of Food Technology, 1979, chap. 6.
2. **Szczesniak, A. S., Loew, B. S., and Skinner, E. Z.,** Consumer texture profile technique, *J. Food Sci.,* 40, 1243, 1975.
3. **Moskowitz, H. R.,** Correlating sensory and instrumental measures in food texture, *Cereal Foods World,* 22, 223, 1979.
4. **Brandt, M. A., Skinner, E. Z., and Coleman, J. A.,** Texture profile method, *J. Food Sci.,* 28(4), 404, 1963.
5. **Szczesniak, A. S., Brandt, M. A., and Friedman, H. H.,** Development of standard rating scales for mechanical parameters of texture and correlation between the objective and the sensory methods of texture evaluation, *J. Food Sci.,* 28(4), 397, 1963.
6. **Szczesniak, A. S.,** Classification of textural characteristics, *J. Food Sci.,* 28, 385, 1963.
7. **Schwartz, N.,** Method to skin care products, *J. Texture Stud.,* 6, 33, 1975.
8. **Moskowitz, H. R.,** Magnitude Estimation: notes on how, what, where and why to use it, *J. Food Qual.,* 1, 195, 1978.
9. **Moskowitz, H. R.,** Application of sensory assessment to food evaluation. II. Methods of ratio scaling, *Lebensmittel-Wissenschaft Technol.,* 8(6), 249, 1975.
10. **Neilson, A. J.,** Time intensity studies, *Drug Cosmet. Ind.,* 80, 452, 1957.
11. **Civille, G. V. and Liska, I. H.,** Modifications and applications to foods of the General Foods sensory texture profile technique, *J. Texture Stud.,* 6, 19, 1975.
12. **Sensory Evaluation Division, Institute of Food Technologists,** Sensory Evaluation Guide for testing food and beverage products, and Guidelines for the preparation and review of papers reporting sensory evaluation data, *Food Technol.,* 35(11), 50, 1981.
13. **Caul, J. F.,** The profile method of flavor analysis, *Adv. Food Res.,* 7, 1, 1957.
14. **Cairncross, S. E. and Sjostrom, L. B.,** Flavor profiles — a new approach to flavor problems, *Food Technol.,* 4, 308, 1950.
15. **Civille, G. V. and Szczesniak, A. S.,** Guidelines to training a texture profile panel, *J. Texture Stud.,* 4, 204, 1973.
16. **Stone, H., Sidel, J., Oliver, S., Woolsey, A., and Singleton, R. C.,** Sensory evaluation by quantitative descriptive analysis, *Food Technol.,* 28(11), 24, 1974
17. **Stone, H. and Sidel, J. L.,** *Sensory Evaluation Practices,* Academic Press, Orlando, Fla., 1985.
18. **Pangborn, R. M., Lewis, M. J., and Yamashita, J. F.,** Comparison of time-intensity with category scaling of bitterness of iso-o-acids in model systems and in beer, *J. Inst. Brew.,* 89, 349, 1983.
19. **Schmitt, D. J., Thompson, L. J., Malek, D. M., and Munroe, J. H.,** An improved method for evaluating time-intensity data, *J. Food Sci.,* 49, 539, 1984.
20. **Guinard, J.-K., Pangborn, R.-M., and Lewis, M. J.,** Effect of repeated ingestion on perception of bitterness in beer, *J. Am. Soc. Brew. Chem.,* in press.
21. **Larson-Powers, N. and Pangborn, R.-M.,** Paired comparison and time-intensity measurements of the sensory properties of beverages and gelatins containing sucrose or synthetic sweeteners, *J. Food Sci.,* 43, 41, 1978.
22. **Guinard, J-X., Pangborn, R.-M., and Shoemaker, C. F.,** Computerized procedure for time-intensity sensory measurements, *J. Food Sci.,* 50, 543, 1985.
23. **Muñoz, A.,** Development and application of texture reference scales, *J. Sensory Stud.,* 1, 55, 1986.
24. **Civille, G. V.,** unpublished.

Chapter 9

AFFECTIVE TESTS:
CONSUMER TESTS AND IN-HOUSE PANEL ACCEPTANCE TESTS

I. PURPOSE AND APPLICATIONS

The primary purpose of affective tests is to assess the personal response (preference and/ or acceptance) by current or potential customers of a product, a product idea, or specific product characteristics.

Affective tests are used mainly by producers of consumer goods, but also by providers of services such as hospitals and banks, and even the Armed Forces, where many tests were first developed (see Chapter 1, Introduction). Each year, consumer tests are used more and more. They have proven highly effective as a principal tool in designing products or services that will sell in larger quantity and/or attract a higher price. The companies that prosper are seen to excel in consumer testing know-how and consequently in knowledge about their consumers.

This chapter gives rough guidelines for the design of consumer tests and in-house panel tests. More detailed discussions are given by Amerine et al.,[1] the ASTM,[2] Moskowitz,[3] Stone and Sidel,[4,5] and Gatchalian.[6] A question that divides these authors is the use of in-house panels for acceptance testing. Our opinion is that this depends on the product: Baron Rothschild doesn't rely on consumer tests for his wines, but Beatrice and Pillsbury need them. For the average company's products, the amounts of testing generated by intended and unavoidable variations in process and raw materials exceed by far the capacity of all the consumer panels in the world, so one has no choice but to use in-house panels for most jobs and then calibrate against consumer tests as often as possible.

Most people today have participated in some form of consumer tests. Typically a test involves 100 to 500 target consumers divided over 3 or 4 cities, say males legal age to 34 who made a purchase of import beer within the last 2 weeks. Potential respondents are screened by phone or in a shopping mall and those selected and willing are given a choice of beers together with a scorecard requesting their preference and reasons therefore along with past buying habits and various demographic questions such as age, income, employment, ethnic background, etc. Results are calculated in the form of preference scores overall and for various subgroups.

The most effective tests for preference or acceptance are based on carefully designed test protocols run among carefully selected subjects with representative products. The choice of test protocol and subjects is based on the project objective. Nowhere in sensory evaluation is the definition of the project objective more critical than with consumer tests which often cost from $10,000 to 50,000 or more. In-house tests are also expensive; the combined cost in salaries and overhead can run $300 to 1500 for a 20-min test involving 20 to 40 people.

From a project perspective, the reasons for conducting consumer tests usually fall into one of the following categories:

- Product maintenance
- Product improvement/optimization
- Development of new products
- Assessment of market potential

A. Product Maintenance

In a typical food or cosmetics company, a large proportion of the product work done by R&D and Marketing deals with the maintenance of current products and their market shares and sales volumes. Research and Development projects may involve cost reduction, substitution of ingredients, process and formulation changes, and packaging modifications, in each case *without* affecting the product's characteristics and overall acceptance. Sensory evaluation tests used in such cases are often difference tests for similarity and/or descriptive tests. However, when a match is not possible it is necessary to take one or more "near misses" out to the consumer, in order to determine if these prototypes will at least achieve parity (in acceptance or preference) with the current product and, perhaps, with the competition.

Product maintenance is a key issue in Quality Control/Quality Assurance and shelf-life/storage projects. Initially it is necessary to establish the "affective status" of the standard or control product with consumers. Once this is done, internal tests can be used to measure the magnitude and type of change over time, condition, production site, raw material sources, etc. with the aid of QC or storage testing. The sensory differences detected by internal tests, large and small, may then be evaluated again by consumer testing in order to determine how large a difference is sufficient to reduce (or increase) the acceptance rating or percent preference vis-a-vis the control or standard.

B. Product Improvement/Optimization

Because of the intense competition among consumer products, companies constantly seek to improve and optimize products, so that they deliver what the consumer is looking for, and thus fare better than the competition. A product improvement project generally seeks to "fix" or upgrade one or two key product attributes, which the consumer has indicated could be improved. A product optimization project typically attempts to manipulate a few ingredient or process variables so as to improve the desired attributes and hence the overall consumer acceptance. Both types of project require the use of a good descriptive panel (1) to verify the initial consumer needs and (2) to document the characteristics of the successful prototype, see Section IV. Examples of projects to improve product attributes are

- Increasing the perceived key flavor attribute, such as, lemon, peanut, coffee, chocolate, etc.
- Increasing an important texture attribute, such as, crisp, moist, etc. or reducing negative properties such as soggy, chalky, etc.
- Improving perceived performance characteristics, such as, longer lasting fragrance, brighter shine, more moisturized skin

In *product improvement,* prototypes are made, tested by the company panel to verify that the desired attribute differences are perceptible, and then tested with consumers to determine the degree of perceived product improvement and its effect on overall acceptance or preference scores.

For *product optimization,*[3,4,7] ingredients or process variables are manipulated; a descriptive analysis identifies the key sensory attributes affected, and consumer tests are conducted to determine if consumers perceive the change in attributes and if such modifications improve the overall ratings.

The study of attribute changes together with consumer scores enables the company to identify and understand those attributes and/or ingredients or process variables that "drive" overall acceptance in the market.

C. Development of New Products

During the typical new product development cycle, affective tests are needed at several critical junctures, e.g., focus groups to evaluate a concept or a prototype; feasibility studies in which the test product is presented to consumers, allowing them to see and touch it; central location tests during product development to confirm that the product's characteristics do confer the expected advantage over the competition; controlled comparisons with the competition during test marketing; renewed comparisons during the reduction-to-practice stage to confirm that the desired characteristics survive into large-scale production; and finally central location and home-use tests during the growth phase to determine the degree of success enjoyed by the competition as it tries to catch up.

Depending on test results at each stage, and the ability of R&D to reformulate or scale up at each step, the new product development cycle can take from a few months to a few years. This process requires the use of several types of affective tests, designed to measure responses to the first concepts, chosen concepts vs. prototypes, different prototypes, and competition vs. prototypes. At any given time during the development process, the test objectives may resemble those of a product maintenance project, for example a pilot plant scale-up, or an optimization project, as described above.

D. Assessment of Market Potential

Typically the assessment of market potential is a function of the Marketing Department, which in turn will consult Sensory Evaluation about the questionnarie and method of testing. Questions about intent to purchase, purchase price, current purchase habits, consumer food habits,[8,9] and the effects of packaging, advertising, and convenience are critical for the acceptance of branded products. The sensory analyst's primary function is to guide research and development. Whether the sensory analyst should also include market oriented questions in consumer testing is a function of the structure of the individual company, including the ability of the Marketing group to provide such data, and the ability of the sensory analyst to assume responsibility for assessing market conditions.

II. THE SUBJECTS/CONSUMERS IN AFFECTIVE TESTS

A. Sampling and Demographics

Whenever a sensory test is conducted, a group of subjects is selected as a *sample* of some larger population, about which the sensory analyst hopes to draw some conclusion. In the case of discrimination tests (difference tests and descriptive tests), the sensory analyst samples individuals with average or above-average abilities to detect differences. It is assumed that if these individuals cannot "see" a difference, the larger human population will be unable to see it. In the case of affective tests, however, it is not sufficient to merely select or sample from the vast human population. Consumer goods and services try to meet the needs of target populations, select markets, or carefully choose segments of the population. Such criteria require that the sensory analyst first determine *the population for whom the product (or service) is intended*, e.g., for a sweetened breakfast cereal, the target population may be children between the ages of 4 and 12, and for a fresh fruit and yogurt blend, the select market may be southern California; for a high-priced jewelry item, or clothing, or an automobile, the segment of the general population may be young, 25 to 35, upwardly mobile professionals, both married and unmarried. Some demographic criteria to be considered in selecting sample subjects are the following.

User group — Based on the rate of consumption of a product by different groups within the population, brand managers often classify users as light, moderate, or heavy users. These terms are highly dependent on the product type and its normal consumption (see Table 1). For specialty products or new products with low incidence in the population, the cost of consumer testing radically increases, because many people must be contacted before the appropriate sample of users can be found.

Age — The ages 4 to 12 are the ones to choose toys, sweets, and cereals; teenagers at 12 to 19 buy clothes and magazines and entertainment; young adults at 20 to 35 receive the most attention in consumer tests (1) because of numbers, (2) because of higher consumption

Table 1
TYPICAL FREQUENCY OF USE OF VARIOUS CONSUMER PRODUCTS

	Product			
User classification	Coffee	Peanut butter, air freshener	Macaroni and cheese	Rug deodorizer
Light	Up to 1 cup/day	1—4 ×/month	Once/2 months	1 ×/year
Moderate	2—5 cups/day	1—6 ×/month	1—4 ×/month	2—4 ×/year
Heavy	5 cups/day	1 × or more/day	Over 2 ×/week	1 ×/month or more

made possible by the absence of family costs, and (3) because lifelong habits and loyalties are formed at this age. Above 35 we buy houses and raise families, and above 65 we use health care and in consumables we tend to look for value for money with an eagle eye. If a product, such as soft drinks, has a broad age appeal, the subjects should be selected by age *in proportion* to their representation in the user population.

Sex — Although women still buy more consumer goods and clothes, and men buy more automobiles, alcohol, and entertainment, the differences in purchasing habits between the sexes continue to diminish. Researchers should use very current figures on users by sex for products such as convenience foods, snacks, personal care products, and wine.

Income — Meaningful groups for most items marketed to the general population are, per family and year:

- Under $20,000
- $20,000 to $40,000
- $40,000 to $70,000
- Over $70,000

Different groups may be relevant at times, e.g., $100,000, $200,000 etc. for yachts over 50 ft.

Geographic location — Because of the regional differences in preference for many products across the U.S., it is often necessary to test products in more than one location and to avoid testing (or use porportional testing) of products for the general population in areas with distinct local preferences, e.g., New York, the Deep South, southern California.

Nationality, region, race, religion, education, employment — These and other factors, such as marital status, number and ages of children in family, pet ownership, size of domicile, etc. may be important for sampling of some products or services. The product researcher, brand manager, or sensory analyst must carefully consider all the parameters which define the target population before choosing the demographics of the sample for a given test.

B. Source of Test Subjects: Employees, Local Residents, the General Population

The need to sample properly from the consuming population, in principle excludes the use of employees and residents local to the company offices, technical center, or plants. However, because of high cost and long turnaround time of consumer tests, companies see a real advantage in using the employees or local population for at least part of their affective testing.

In situations where the project objective is product maintenance (see Section I.A), employees and local residents do not represent a great risk as the test group. In a project oriented to maintaining "sensory integrity" of a current product, employees or local residents familiar with the characteristics of the product can render evaluations which are a good measure of the reactions of regular users. In this case, the employee or local resident judges the relative difference in acceptability or preference of a test sample vis-a-vis the well-known standard or control.

Employee acceptance tests can be a valuable resource when used correctly and when limited to maintenance situations. Because of their familiarity with the product and with testing, employees can handle more samples at a time and can give better discrimination, faster replies, and cheaper service. Employee acceptance tests can be carried out at work in a laboratory, in the style of a central location test, or the employees may take the product home.

However, for new product development, or product optimization, or product improvement, employees or local residents should not be used to represent the consumer. The following are some examples of biases which may result from conducting affective tests with employees:

1. Employees tend to find reasons to prefer the products which they and their fellow employees helped to make, or if morale is bad, find reasons to reject such products. It is therefore imperative that products be disguised. If this is not possible, a consumer panel must be used.
2. Employees may be unable to weigh desirable characteristics against undesirable ones in the same way a consumer would. For example, employees may know that a recent change was made in the process to produce a paler color, and this will make them prefer the paler product and give too little weight to other characteristics. Again, in such a case the color must be disguised, or if this is not possible, outside testing must be used.
3. Where a company makes separate products for different markets, outside tests will be distributed to the target population, but this cannot be done with employees. The way out may be to tell the employee that the product is destined for X market, but sometimes this cannot be done without violating the requirement that the test be blind. If so, recourse must again be had to outside testing.

In summary, the test organizer must plan the test imaginatively and must be aware of every conceivable source of bias. In addition, validity of response must be assured by frequent comparisons with real consumer tests on the same samples. In this way, the organizer and the employee panel members slowly build up knowledge of what the market requires, and this in turn makes it easier to guage the pitfalls and avoid them.

III. CHOICE OF TEST LOCATION

The test location or test site has numerous effects on the results, not only because of the geographic location, but because the place in which the test is conducted defines several other aspects of the way the product is sampled and perceived. It is possible to get different results from different test sites with a given set of samples and consumers. These differences occur as a result of differences in

1. The length of time the products are used/tested.
2. Controlled preparation vs. normal-use preparation of the product.
3. The perception of the product alone in a central location vs. in conjunction with other foods or personal care items in the home.
4. The influence of family members on each other in the home.
5. The length and complexity of the questionnaire.

A. Laboratory Tests

The advantages of laboratory tests are

1.   Product preparation and presentation can be carefully controlled.
2.   Employees can be contacted on short notice to participate.
3.   Color and other visual aspects which may not be fully under control in a prototype, can be masked so that subjects can concentrate on the flavor or texture differences under test.

The disadvantages of laboratory tests are

1.   The locations suggest that the test products originate in the company or specific plant, which may influence biases and expectations because of previous experience.
2.   The lack of normal consumption (e.g., sip test rather than drinking) may influence the detection or evaluation of positive or negative attributes.
3.   Product tolerances in preparation or use may be different from those of home use (e.g., the product may lose integrity under normal use).

B. Central Location Tests

Central location tests are usually conducted in an area where many potential purchasers congregate or can be assembled. The organizer sets up a booth or rents a room at a fair or shopping mall. A product used by schoolchildren may be tested in the school playground, a product for analytical chemists at a professional convention. Respondents are intercepted and screened in the open and those selected for testing are led to a closed-off area. Subjects can also be prescreened by phone and invited to a test site. Typically 50 to 300 responses are collected per location. Products are prepared out of sight and served on uniform plates (cups, glasses) labeled with three-digit codes. The potential for distraction may be high, so instructions and questions should be clear and concise; examples of scoresheets are given in Section IV. In a variant of the procedure, products are dispensed openly from original packaging and respondents are shown storyboards with examples of advertising and how products will be positioned in the market.

The advantages of central location tests are

1.   Respondents evaluate the product under conditions controlled by the organizer; any misunderstandings can be cleared up and a truer response obtained.
2.   The products are tested by the end users themselves which assures the validity of the results.
3.   Conditions are favorable for a high percentage return of responses from a large sample population.

The main disadvantages of central location tests are

1.   The product is being tested under conditions which are quite artificial compared to normal use at home or at parties, restaurants, etc. in terms of preparation, amount consumed, and length and time of use.
2.   The number of questions that can be asked may be quite limited. This in turn limits the information obtainable from the data with regard to the preferences of different age groups, socioeconomic groups, etc.

C. Home-Use Tests

In most cases, home-use tests (or home placement tests) represent the ultimate in consumer tests. The product is tested under its normal conditions of use. The participants are selected to represent the target population. The entire family's opinion is obtained, and the influence of one family member on another is taken into account. In addition to the product itself the home-use test provides a check on the package to be used and the product preparation instructions, if used. Typical panel sizes are 75 to 300 per city in 3 or 4 cities. Generally two products are compared. The first is used for 4 to 7 days and the scoresheet filled in, after which the second is supplied and rated. The two products should not be provided together because of the opportunities for using the wrong clues as the basis for evaluation, or assigning responses to the wrong scoresheet. Examples of scoresheets are given in Sections IV and V.

The advantages of home-use tests are[3]

1. The product is prepared and consumed under natural conditions of use.
2. Information regarding preference between products will be based on stabilized (from repeated use) rather than on first impressions alone as in a mall intercept test.
3. Cumulative effect on the respondent from repeated use can provide information about the potentials for repeat sale.
4. Statistical sampling plans can be fully utilized.
5. Because more time is available for the completion of the scoresheet, more information can be collected regarding the consumer's attitudes to various characteristics of the product, including sensory attributes, packaging, price, etc.

The disadvantages of the home use test are

1. A home use test is time consuming, taking from 1 to 4 weeks to complete.
2. It can cater to a much smaller set of respondents than a central location test; to reach many residences would be unnecessarily lengthy and expensive.
3. The possibility of nonresponse is greater; unless frequently reminded, respondents forget their tasks; haphazard responses may be given as the test draws to a close.
4. A maximum of three samples can be compared; any larger number will upset the natural use situation which was the reason for choosing a home-use test in the first place.
5. The tolerance of the product for mistakes in preparation is tested. The resulting variability in preparation along with variability from the time of use, and from other foods or products used with the test product, combine to produce a large variability across a relatively small sample of subjects.

IV. METHODS USED IN AFFECTIVE TESTS/CONSUMER TESTS

A. Primary Response: Preference or Acceptance?

Affective tests can be classified into two main categories on the basis of the primary task of the test:

Task	Test type	Questions
Choice	Preference tests	Which sample do you prefer?
		Which sample do you like better?
Rating	Acceptance tests	How much do you like the product?
		How acceptable is the product?

In addition to these questions, which can be asked in several ways using various response forms (see below), the test design often asks secondary questions about the reasons for the expressed preference or acceptance (see Section IV.B below on Attribute Diagnostics).

1. Preference Tests

The choice of preference or acceptance for a given affective test should be based again on the project objective. If the project is specifically designed to pit one product *directly* against another in situations such as *product improvement* or *parity with competition,* then a preference test is indicated. The preference test forces a choice of one item over another or others. What it does not do is indicate whether any of the products are liked or disliked. Therefore, the sensory researcher must have prior knowledge of the "affective status" of the current product or competitive product, against which he is testing.

Preference tests can be classified as follows:

Test type	No. of samples	Preference
Paired preference	2	A choice of one sample over another (A-B)
Rank preference	3 or more	A relative order of preference of samples (A-B-C-D)
Multiple paired Preference (all pairs)	3 or more	A series of paired samples with all samples paired with all others (A-B, A-C, A-D, B-C, B-D, C-D)
Multiple paired Preference (selected pairs)	3 or more	A series of paired samples with one or two select samples (e.g. control) paired with two or more others (not paired with each other) (A-C, A-D, A-E, B-C, B-D, B-E)

See Chapter 6, Section III, for a discussion of principles, procedures, and analysis of paired and multipaired tests.

a. Example 1. Paired Preference. Improved Peanut Butter

Problem/situation — In response to consumer requests for a product "with better flavor with more peanutty character", a product improvement project has yielded a prototype which was rated significantly more peanutty in an attribute difference test (such as discussed in Chapter 6, Section III.A). Marketing wishes to confirm that the prototype is indeed preferred to the current product, which is enjoying large volume sales.

Test objective — To determine whether the prototype is preferred over the current product.

Test design — This test is one-sided as the prototype was developed to be more peanutty in response to consumer requests. A group of 100 subjects, prescreened as users of peanut butter, are selected and invited to a central location site where they receive the two samples in simultaneous presentation, half in the order A-B, the other half B-A. All samples are coded with three-digit random numbers. Subjects are encouraged to make a choice (see discussion on forced choice, Chapter 6, Section II.A.2). The scoresheet is shown in Figure 1. The null hypothesis is H_o: The preference for the higher-peanut flavor prototype $\leq 50\%$. The alternative hypothesis is H_a: The preference for the prototype $> 50\%$.

Screen samples — Samples used are those already subjected to the attribute difference test described above, in which a higher level of peanut flavor was confirmed.

Conduct test — The method described in Chapter 6, Section III.A was used; 62 subjects preferred the prototype. It is concluded from Table T8 that a significant preference exists for the prototype over the current product.

Interpret results — The new product can be marketed in place of the current with a label stating: More Peanut Flavor.

2. Acceptance Tests

When a product researcher needs to determine the "affective status" of a product, i.e., how well it is liked by consumers, an acceptance test is the correct choice. The product is compared to a well-liked company product or that of a competitor, and a hedonic scale,

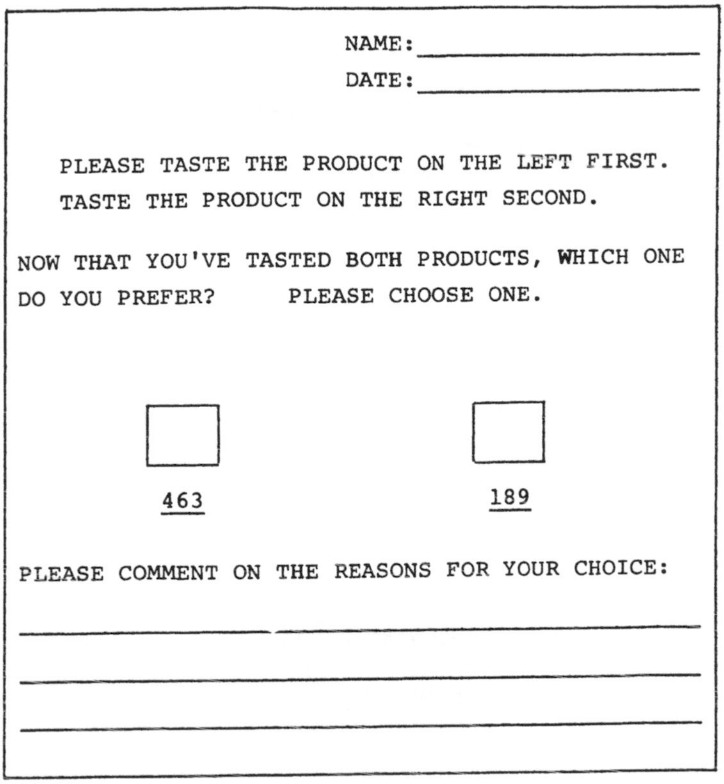

```
                    NAME: _____
                    DATE: _____


    PLEASE TASTE THE PRODUCT ON THE LEFT FIRST.
    TASTE THE PRODUCT ON THE RIGHT SECOND.

  NOW THAT YOU'VE TASTED BOTH PRODUCTS, WHICH ONE
  DO YOU PREFER?     PLEASE CHOOSE ONE.



            ┌─────┐              ┌─────┐
            │     │              │     │
            └─────┘              └─────┘
              463                  189


  PLEASE COMMENT ON THE REASONS FOR YOUR CHOICE:

  _____  _____

  _____

  _____
```

FIGURE 1.   Scoresheet for Example 1, paired preference test, improved peanut butter.

Table 2
VERBAL HEDONIC
SCALE, USED IN
ACCEPTANCE TESTS

0	Like extremely
0	Like very much
0	Like moderately
0	Like slightly
0	Neither like nor dislike
0	Dislike slightly
0	Dislike moderately
0	Dislike very much
0	Dislike extremely

such as those shown in Tables 2 to 4 and in Figure 2, is used to indicate degrees of unacceptable to acceptable, or dislike to like.

From relative acceptance scores one can infer preference, the sample with the higher score is preferred. The best (more discriminating, more actionable) results are obtained with scales that are balanced, i.e., have an equal number of positive and negative categories, and have steps of equal size. The scales shown in Table 5 are not as widely used because they are unbalanced, unevenly spaced, or both. Scale G in Table 5 for example, is heavily loaded with positive (good to excellent) categories, and the space between Poor and Fair is clearly

Table 3
NUMERICAL HEDONIC
SCALE, USED IN
ACCEPTANCE TESTS

Rate the Product Using the Scale below
which Ranges from 1 to 11

1	——————————	
2	——————————	
3	——————————	Dislike
4	——————————	
5	——————————	
6	——————————	
7	——————————	
8	——————————	
9	——————————	Like
10	——————————	
11	——————————	

Table 4
PURCHASE INTENT
SCALE, USED IN
ACCEPTANCE TESTS

0	Definitely would buy
0	Probably would buy
0	Maybe/maybe not
0	Probably would not buy
0	Definitely would not buy

larger than that between Extremely Good and Excellent. The difference between the latter may be unclear to many people. Acceptance tests are in fact very similar to attribute difference tests (see Chapter 6, Section II) except that the attribute here is *acceptance* or *liking*. Different types of scales: category (as shown in Tables 2 to 5 and Figure 2) line scales, or ME scales can be used to measure the degree of liking for a product.

a. Example 2: Acceptance of Two Prototypes Relative to a Competitive Product. High Fiber Breakfast Cereal

Problem/situation — A major cereal manufacturer has decided to enter the higher fiber cereal market and has prepared two prototypes. Another major cereal producer already has a brand on the market that continues to grow in market share and leads among the high fiber brands. The researcher needs to obtain acceptability ratings for his two prototypes compared to the leading brand.

Project objective — To determine whether one or the other prototype enjoys sufficient acceptance to be test marketed against the leading brand.

Test objective — To measure the acceptability of the two prototypes and the market leader among users of high fiber cereals.

Screen the samples — During a product review, several researchers, brand marketing staff, and the sensory analyst taste the prototypes and competitive cereal which are to be submitted to a home placement test.

Test design — Each prototype is paired with the competitor in a separate sequential evaluation, in which each product is used for 1 week. The prototypes and the competitive

FIGURE 2. Facial hedonic scale, used in acceptance tests.

product are each evaluated first in half of the test homes. Each of the 150 qualified subjects is asked to rate the products on the nine point verbally anchored hedonic scale shown in Figure 2.

Conduct test — One product (prototype or competition) is placed in the home of each prescreened subject for 1 week. After the questionnaire is filled in and the first product removed, the second product is given to the subject to use for the second week. The second questionnaire and remaining samples are collected at the end of the second week.

Analyze results — Separate paired *t*-tests (see Chapter 11) are conducted for each prototype vs. the competition. The mean acceptability scores of the samples were

	Prototype	Competition	Difference
Prototype 1	6.6	7.0	−0.4
Prototype 2	7.0	6.9	+0.1

The average difference between prototype 1 and the competition was significantly different from zero, that is, the average acceptability of prototype 1 is significantly less than the competition. There was no significant difference between prototype 2 and the competition.

Interpret results — The project manager concludes that Prototype 2 did as well as the competition, and the group recommends it as the company's entry into the high fiber cereal field.

B. Assessment of Individual Attributes

As part of a consumer test, researchers often endeavor to determine the *reasons* for any preference or rejection by asking additional questions about the sensory attributes (appearance, aroma/fragrance, sound, flavor, texture/feel). Such questions can be classified into the following groups:

Table 5

EXAMPLES OF HEDONIC SCALES THAT ARE
UNCLEAR IN BALANCE OR SPACING

Nine-point wonderful
  Think it's wonderful
  Like it very much
  Like it quite a bit
  Like it slightly
  Neither like nor dislike it
  Dislike it slightly
  Dislike it quite a bit
  Dislike it very much
  Think it's terrible

Seven-point wonderful
  Like it very much
  Like it quite a bit
  Like it slightly
  Neither like nor dislike it
  Dislike it slightly
  Dislike it quite a bit
  Dislike it very much

Seven-point excellent
  Excellent
  Very good
  Good
  Fair
  Poor
  Very poor
  Terrible

Five-point excellent
  Excellent
  Good
  Fair
  Poor
  Terrible

Nine-point quartermaster (unbal.)
  Like extremely
  Like strongly
  Like very well
  Like fairly well
  Like moderately
  Like slightly
  Dislike slightly
  Dislike moderately
  Dislike intensely

Six-point wonderful (unbal.)
  Wonderful, think it's great
  I like it very much
  I like it somewhat
  So-so, it's just fair
  I don't particularly like it
  I don't like it at all

Six-point excellent (unbal.)
  Excellent
  Extremely good
  Very good
  Good
  Fair
  Poor

1. Affective responses to
      attributes:
    Preference: Which sample do you prefer for fragrance?
    Hedonic: How do you like the texture of this product?
                           [Like extremely -------------------- Dislike extremely]

2. Intensity response to
      attribute:
    Strength: How strong/intense is the crispness of this cracker?
                           [None ------------------------------------- Very strong]

3. Appropriateness of
      intensity:
    Just right: Rate the sweetness of this cereal:
                           [Not at all sweet enough ------------ Much too sweet]

Table 6
ATTRIBUTE DIAGNOSTICS. EXAMPLE OF
ATTRIBUTE-BY-PREFERENCE QUESTIONS

1. Which sample did you prefer overall?   467____   813____
2. Which did you prefer for color?   467____   813____
3. Which did you prefer for cola impact?   467____   813____
4. Which did you prefer for citrus flavor?   467____   813____
5. Which did you prefer for spicy flavor?   467____   813____
6. Which did you prefer for sweetness?   467____   813____
7. Which did you prefer for body?   467____   813____

Table 7
ATTRIBUTE DIAGNOSTICS QUESTIONNAIRE WITH A SINGLE SAMPLE,
USING HEDONIC RATING OF EACH ATTRIBUTE

☐ Like extremely
☐ Like very much
☐ Like moderately
☐ Like slightly
☐ Neither like not dislike
☐ Dislike slightly
☐ Dislike moderately
☐ Dislike very much
☐ Dislike extremely

Using the above scale rate the following:
[Scale could be repeated after each question]
How do you feel *overall* about this beverage?_____
How do you feel about the color?_____
How to you feel about the cola impact?_____
How do you feel about the citrus flavor?_____
How do you feel about the spice flavor?_____
How do you feel about the sweetness?_____
How do you feel about the body?_____

Tables 6 to 8 are examples of questionnaires containing various types of attribute questions. In Table 6, a preference questionnaire with two samples, respondents are asked, for each attribute, which sample they prefer. In Table 7, an "attribute diagnostics" questionnaire with a single sample, respondents rate each attribute on a scale from "like extremely" to "dislike extremely". Such questionnaires are considered less effective in determining the importance of each attribute, because subjects often rate the attributes similar to the overall response, and the result is a series of attributes which have a "halo" of the general response. In addition, if one attribute does receive a negative rating, the researcher has no way of determining the direction of the dislike. If a product's texture is disliked, is it "too hard" or "too soft"? "too thick" or "too thin"?

The "just right" scales shown in Table 8 and Figure 3 allow the researcher to assess the intensity of an attribute relative to some mental criterion of the subjects. Just right scales cannot be analyzed by calculating the mean response, as the scale might be unbalanced or unevenly spaced, depending on the relative intensities and appropriateness of each attribute in the mind of the consumer. The following procedure is recommended:

1. Calculate the percentage of subjects who respond in each category of the attribute. Example:

Table 8
ATTRIBUTE DIAGNOSTICS. "JUST RIGHT" SCALES

1. *Overall* How do you feel about this product?

2. The color of this beverage is:
   _____Very light
   _____Somewhat light
   _____Just right
   _____Somewhat dark
   _____Very dark

3. This beverage has:
   _____Much too little cola impact
   _____Somewhat too little cola impact
   _____Just right
   _____Somewhat too much cola impact
   _____Too much cola impact

4. The citrus flavor of this beverage is:
   _____Much too little citrus
   _____Somewhat too little citrus
   _____Just right
   _____Somewhat too much citrus
   _____Much too much citrus

5. This beverage is:
   _____Much too bland
   _____Somewhat too bland
   _____Just right
   _____Somewhat too spicy
   _____Much too spicy

6. This beverage is:
   _____Not at all sweet enough
   _____Somewhat not sweet enough
   _____Just right
   _____Somewhat too sweet
   _____Much too sweet

7. The body of this beverage is
   _____Much too thin
   _____Somewhat too thin
   _____Just right
   _____Somewhat too thick
   _____Much too thick

1. THIS BEVERAGE IS:

   MUCH TOO LIGHT IN COLOR ☐ ☐ ☐ ☐ ☐ ☐ MUCH TOO DARK IN COLOR

2. THIS BEVERAGE HAS:

   TOO LITTLE COLA IMPACT ☐ ☐ ☐ ☐ ☐ ☐ TOO MUCH COLA IMPACT

3. THIS BEVERAGE HAS:

   TOO LITTLE CITRUS FLAVOR ☐ ☐ ☐ ☐ ☐ ☐ TOO MUCH CITRUS FLAVOR

4. THIS BEVERAGE IS:

   TOO BLAND ☐ ☐ ☐ ☐ ☐ ☐ TOO SPICY

5. THIS BEVERAGE IS:

   NOT AT ALL SWEET ENOUGH ☐ ☐ ☐ ☐ ☐ ☐ MUCH TOO SWEET

6. THE BODY OF THIS BEVERAGE IS:

   MUCH TOO THIN ☐ ☐ ☐ ☐ ☐ ☐ MUCH TOO THICK

7. OVERALL HOW DID YOU FEEL ABOUT THIS PRODUCT?

   MUCH TOO POOR ☐ ☐ ☐ ☐ ☐ ☐ EXCELLENT

FIGURE 3.   Attribute Diagnostics. Implied "just right" scales.

% Response	5	15	40	25	15
Category	Much too little	Somewhat too little	Just right	Somewhat too much	Much too much

2. Using a χ^2 test (Chapter 11) compare the distribution of responses to that obtained by a successful brand.

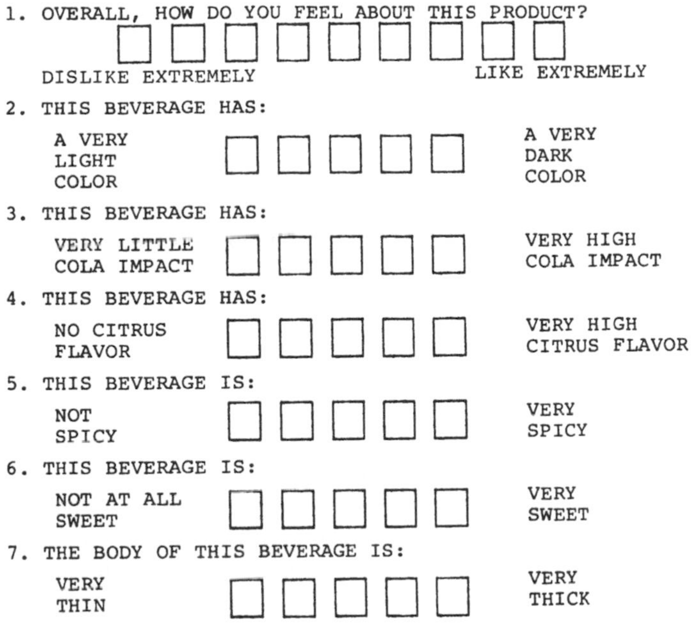

1. OVERALL, HOW DO YOU FEEL ABOUT THIS PRODUCT?

☐☐☐☐☐☐☐☐

DISLIKE EXTREMELY   LIKE EXTREMELY

2. THIS BEVERAGE HAS:

A VERY LIGHT COLOR   ☐☐☐☐☐   A VERY DARK COLOR

3. THIS BEVERAGE HAS:

VERY LITTLE COLA IMPACT   ☐☐☐☐☐   VERY HIGH COLA IMPACT

4. THIS BEVERAGE HAS:

NO CITRUS FLAVOR   ☐☐☐☐☐   VERY HIGH CITRUS FLAVOR

5. THIS BEVERAGE IS:

NOT SPICY   ☐☐☐☐☐   VERY SPICY

6. THIS BEVERAGE IS:

NOT AT ALL SWEET   ☐☐☐☐☐   VERY SWEET

7. THE BODY OF THIS BEVERAGE IS:

VERY THIN   ☐☐☐☐☐   VERY THICK

FIGURE 4. Attribute Diagnostics. Use of simple intensity scales for each attribute.

A similar approach is to use an intensity scale (without midpoint) for each attribute, see Figure 4. In order to assess how appropriate each of these attributes is, the intensity values must be related to the attribute values for the consumer's "ideal" product. The studies done by General Foods on the Consumer Texture Profile[11] method show high correlations between acceptance ratings and the degree to which various products approach the consumer's ideal.

V. DESIGN OF QUESTIONNAIRES

In designing questionnaires for affective testing the following guidelines are recommended:

1. Keep the length of the questionnaire in proportion to the amount of time the subject *expects* to be in the test situation. Subjects can be contracted to spend hours testing several products with extensive questionnaires. At the other extreme, a few questions may be enough information for some projects. Design the questionnaire to ask the minimum number of questions to achieve the project objective, then set up the test so that the respondents expect to be available for the appropriate time span.
2. Keep the questions clear and somewhat similar in style. Use the same type of scale, whether preference, hedonic, just right, or intensity scale, for all attributes. Have the scale go in the same direction, e.g., [Too little-Too much] for each attribute, so that the subject does not have to stop and decode each question.
3. Direct the questions to address the primary differences between/among the products in the test. Attribute questions should relate to the attributes which are detectable in the products and which differentiate among them. Subjects will not give clear answers to questions about attributes they cannot perceive, or differences they cannot detect.
4. Use only questions which are actionable. Do not ask questions to provide data for which there is no appropriate action. If one asks subjects to rate the attractiveness of a package, and the answer comes back that the package is somewhat unattractive, does the researcher know what to "fix" or change to alter that rating?

5.   Always provide one or two spaces on a scoresheet for open-ended questions. For example, ask why a subject responded the way he/she did to a preference or acceptance question.

6.   Place the overall question for preference or acceptance in the place on the scoresheet which will elicit the most considered response. In many cases the overall acceptance is of primary importance, and analysts rightly tend to place it first on the scoresheet. However, in cases where a consumer is asked several specific questions about appearance and/or aroma before the actual consumption of the product, it is necessary to wait until those attributes are evaluated and rated before addressing the total acceptance or preference question.

VI. USING OTHER SENSORY METHODS TO SUPPLEMENT AFFECTIVE TESTING

A. Relating Affective and Descriptive Data

Product development professionals handling both the Research and Development and marketing aspects of a product cycle recognize that the consumer's response in terms of overall acceptance and purchase intent is the bottom line in the decision to go or not go with a product or concept.

Despite the recognition of the need for affective data, the product development team is generally unsure about what the consumer means when asked about actual sensory perceptions. When a consumer rates a product as too dry or not chocolatey enough, is he really responding to perceived moistness/dryness or perceived chocolate flavor, or is he responding to words that are associated in his mind with goodness or badness in the product? Too many researchers are taking the consumer's response at face value — as the researcher uses the sensory terms — and these researchers end up fixing attributes that may not be broken.

One key to decoding consumer diagnostics and consumer acceptance is to measure the perceived sensory properties of a product using a more objective sensory tool. The trained descriptive or expert panel provides a thumbprint or spectrum of a product's sensory properties. This sensory documentation constitutes a list of real attribute characteristics or differences among products which can be used both to design relevant questionnaires and to interpret the resulting consumer data after the test is completed. By associating consumer data with panel data and when possible with ingredient and processing variables, or with instrumental or chemical analyses, the researcher can discover the relationships between the product's attributes and the ultimate bottom line, consumer acceptance.

B. Using Affective Data to Define Shelf-Life or Quality Limits

In Chapter 8, Section VI, we described a "modified" or short-version descriptive procedure whose principal use is to define QA/QC or shelf-life limits. In a typical case, the first step is to send the fresh product out for an acceptability test in a typical user group. This initial questionnaire may contain additional questions asking the consumer to rate a few important attributes.

The product is also rated for acceptability and key attributes by the modified panel, and this evaluation is repeated at regular intervals during the shelf storage period, each time comparing the stored product with a control, which may be the same product stored under conditions that inhibit perceptible deterioration (e.g., deep freeze storage under nitrogen) or if this is not possible, fresh product of current production.

When a significant difference is found by the modified panel, in overall difference from the control and/or in some major attribute(s), the samples are sent again to the user group to determine if the statistically significant difference is meaningful to the consumer. This is repeated as the difference grows with time of shelf storage. Once the size of a panel difference can be related to what reduces consumer acceptance or preference, the internal panel can be used in future to monitor regular production in shelf-life studies, with assurance that the results are predictive of consumer reaction.

1. Example 5: Shelf-Life of Sesame Cracker

Problem/situation — A company wishes to define the shelf-life of a new sesame cracker in terms of the "sell by" date which will be printed on packages on the day of production.

Project objective — To determine at what point during shelf storage the product will be considered "off", "stale", or "not fresh" by the consumer.

Test objective — (1) Using a research panel trained for the purpose of determining the key attributes of the product at various points during shelf storage. (2) Submitting the product to consumer acceptance tests (a) initially, (b) when the research panel first establishes a difference, and (c) at intervals thereafter, until the consumers establish a difference.

Test design — Samples of a single batch of the sesame crackers were held for 2, 4, 6, 8, and 12 weeks under four different sets of conditions: "control" = near freezing in airtight containers; "ambient" = 70°F/50% RH; "humid" = 85°F/70% RH; and "hot" = 100°F/30% RH.

Subjects — Panelists (25) from the R&D lab are selected for ability to recognize the aromatics of stale sesame crackers, i.e., the cardboard aromatic of the stale base cracker and the painty aromatic of oxidized oil from the seeds. Consumers (250) must be users of snack crackers and are chosen demographically to represent the target population.

Sensory methods — The research panel used the questionnaire in Figure 5 and was trained to score the test samples on the seven line scales, which represent key attributes of appearance, flavor, and texture related to the shelf-life of crackers and sesame seeds. Research panelists also received a sample marked "control" with instructions to use the last line of the form as a difference-from-control test, see Chapter 6, Section II.F. The panelists were informed that test samples were part of a shelf-life study, and that occasional test samples would consist of a blind control plus freshly prepared "control product" (such information reduces the tendency of panelists in shelf-life testing to anticipate more and more degradation in products).

The consumers on each occasion received two successive coded samples (the test product and the control, in random order), each with the scoresheet in Figure 6 which they filled in on the spot and returned to the interviewer.

Analyze results — The initial acceptance test, in which the 250 consumers received two fresh samples, provided a baseline rating of 7.2 for both, and the accompanying attribute ratings indicated that the crackers were perceived fresh and crisp.

The same two identical samples were rated 3.2 (out of 15) on the difference-from-control scale by the research panel. The 2- and 4-week samples showed no significant differences. At the 6-week point, the "humid" sample received a difference-from-control rating of 5.9, which was significantly different from 3.2. In addition, the "humid" sample was rated 4.2 in cardboard flavor (against 0 for the fresh control) and 5.1 in crispness (against 8.3 for the fresh control), both significant differences by ANOVA.

The 6-week "humid" samples were then tested by the consumers and were rated 6.7 on acceptance, against 7.1 for the control ($p < 0.05$). The rating for "fresh toasted flavor" also showed a significant drop.

The product researcher decided to conduct consumer tests with the other two test samples ("ambient" and "hot" as soon as the difference-from-control ratings by the research panel exceeded 5.0. Subsequent tests showed that consumers were only sensitive to differences which were rated 5.5 or above by the research panel. All further shelf-life testing on sesame crackers used the 5.5 difference-from-control rating as the critical point above which differences were not only statistically significant, but potentially meaningful to the consumer.

```
┌─────────────────────────────────────────────────────────────────┐
│                  EVALUATION OF SESAME CRACKER                     │
├─────────────────────────────────────────────────────────────────┤
│                                                                   │
│  INSTRUCTIONS                                                     │
│  1.  Evaluate the cracker for appearance, flavor and tex-        │
│      ture by placing a mark on each line below:                  │
│  Appearance                                                      │
│  Surface color  |──────────────────────────────────────────|    │
│                 light                                    dark    │
│  Flavor                                                          │
│  Toasted wheat  |──────────────────────────────────────────|    │
│                 none                                   strong    │
│  Sesame seed    |──────────────────────────────────────────|    │
│                 none                                   strong    │
│  Cardboard      |──────────────────────────────────────────|    │
│                 none                                   strong    │
│  Painty         |──────────────────────────────────────────|    │
│                 none                                   strong    │
│  Texture                                                        │
│  Hardness       |──────────────────────────────────────────|    │
│                 soft                                     hard    │
│  Crispness      |──────────────────────────────────────────|    │
│                 soggy                                   crisp    │
│                                                                   │
│  2.  Compare the cracker with the control and indicate           │
│      the amount of difference between them by placing a          │
│      mark on the line below:                                     │
│                                                                   │
│            |──────────────────────────────────────────|         │
│        no difference                          very different    │
├─────────────────────────────────────────────────────────────────┤
│  Comments  _____          │
│            _____          │
│            _____          │
├─────────────────────────────────────────────────────────────────┤
│  Name  _____  Date  _____         │
└─────────────────────────────────────────────────────────────────┘
```

FIGURE 5.   Research panel scoresheet for Example 5, shelf-life of sesame cracker.

EVALUATION OF SESAME CRACKER

INSTRUCTIONS

1. Overall evaluation. Place a mark in the box which you
    feel best describes how you like this product:

Like extre-mely	Like very much	Like mode-rate-	Like slig-htly	Neither like nor dislike	Dislike slig-htly	Dislike mode-rately	Dislike very much	Dislike extre-mely

2. Indicate by placing a mark how you feel the product
    rates in each category below:

Appearance

Color

    light   dark

Flavor

Salty

    not at all salty   very salty

Sesame flavor

    no sesame flavor   strong flavor

Fresh toasted flavor

    stale/not fresh   very fresh

Texture

Crispness

    soggy   crisp

Aftertaste

    unpleasant   pleasant

Comments _____

_____

_____

Name _____   Date _____

FIGURE 6. Consumer scoresheet for Example 5, shelf-life of sesame cracker.

REFERENCES

1. **Amerine, M. A., Pangborn, R. M., and Roessler, E. B.,** *Principles of Sensory Evaluation of Food,* Academic Press, New York, 1965, chap. 9.
2. **Schaefer, E. E., Ed.,** *ASTM Manual on Consumer Sensory Evaluation,* Special Technical Publication 682, American Society for Testing and Materials, Philadelphia, 1979.
3. **Moskowitz, H. R.,** *Product Testing and Sensory Evaluation of Foods. Marketing and R&D Approaches,* Food & Nutrition Press, Westport, Conn., 1983.
4. **Sidel, J. L. and Stone, H.,** Consumer testing considerations, in *Sensory Evaluation Methods for the Practicing Food Technologist,* Johnson, M. R., Ed., Institute of Food Technologists, Chicago, 1979, 1.
5. **Stone, H. and Sidel, J. L.,** *Sensory Evaluation Practices,* Academic Press, Orlando, Fla., 1985, 227.
6. **Gatchalian, M. M.,** *Sensory Evaluation Methods with Statistical Analysis,* College of Home Economics, University of the Philippines, Diliman, Quezon City, 1981, 230.
7. Institute of Food Technologists, *Sensory Evaluation Short Course,* Institute of Food Technologists, Chicago, 1979.
8. **Meiselman, H. L.,** Consumer studies of food habits, in *Sensory Analysis of Foods,* Piggott, J. R., Ed., Elsevier, Amsterdam, 1984, chap. 8.
9. **Barker, L.,** *The Psychobiology of Human Food Selection,* AVI, Westport, Conn., 1982.
10. **Gardner, D.,** Statistics, Lecture, Sensory Evaluation Course, Civille, G. V., Course Director, The Center for Professional Advancement, East Brunswick, N.J., July 1985.
11. **Szczesniak, A. S., Skinner, E. Z., and Loew, B. J.,** Consumer texture profile method, *J. Food Sci.,* 40, 1253, 1975.

Chapter 10

SELECTION AND TRAINING OF PANEL MEMBERS

I. INTRODUCTION

This section is partly based on ASTM Special Technical Publication 758, "Guidelines for the Selection and Training of Sensory Panel Members"[1] and on the ISO's "Guide for Selection and Training of Assessors".[2] The development of a sensory panel deserves thought and planning with respect to the inherent need for the panel, the support from the organization and its management, the availability and interest of panel candidates, the need for screening of training samples and references, and the availability and condition of the panel room and booths. In the food, fragrance, and cosmetic industries, the sensory panel is the company's single most important tool in research and development and in quality control. The success or failure of the panel development process depends on the strict criteria and procedures used to select and train the panel.

The project objective of any given sensory problem or situation determines the criteria for selection and training of the subjects. Too often in the past,[3] the sole criterion was a low threshold for one or more of the basic tastes. Today's sensory analyst used a wide selection of tests, specifically selected to correspond to the proposed training regimen and end-use of the panel. Taste acuity is only one aspect; much more important is the ability to discern and describe a particular sensory characteristic in a "sea" or "fog" of other sensory impressions.

This chapter describes specific procedures for the decision to establish a panel, the selection and training of both difference and descriptive panels, and ways to monitor and motivate panels. This chapter does not apply to consumer testing (see Chapter 9) which used naive subjects representative of the consuming population. Although the text uses the language of a commercial organization which exists to develop, manufacture, and sell a "product" and has its "upper" and "middle management" and reward structure, the system described can be easily modified to fit the needs of other types of organization such as universities, hospitals, civil or military service organizations, etc.

II. PANEL DEVELOPMENT

Before a panel can be selected and trained, the sensory analyst must establish that a need exists in the organization and that commitment can be obtained to expend the required time and money to develop a sensory tool.[4] Upper management and the project group (R&D or QA/QC) must see the *need* to make decisions based on sound sensory data with respect to overall differences and attribute differences (difference panels) or full descriptions of product standards, product changes over time or ingredient and processing manipulation, and for construction and interpretation of consumer questionnaires (descriptive panels). The sensory analyst must also define the resources required to develop and maintain a sensory panel system.

Personnel — Heading the list of resources required is (1) a large enough pool of available candidates from which the panel can be selected, (2) a sensory staff to implement the selection, training, and maintenance procedures, and a qualified person to conduct the training process. Panelists most often come from within the organization as they are located at the site where the samples are prepared, e.g., R&D facility or plant. However, some companies choose to test products at a different site, which may be another company facility, or they use outside panelists recruited from the neighboring community. Panel candidates and management must understand in advance the amount of time required (personnel hours) for the selection and training of the particular panel in question. An assessment of the number of hours needed for panelists, technicians, and panel leader should be presented and accepted before the development process is initiated. The individual designated to select and train the panel is often a member of the sensory staff who is experienced and trained in the specific selection and training techniques needed for the problem at hand.

Facilities — The physical area for the selection, training, and ongoing work of a panel must be defined before development of the panel begins. A training room and panel testing facilities (booths and/or round table/conference room) must have the proper environmental controls (see Chapter 3), must be of sufficient size to handle all of the panelists and products projected and must be located near to the product preparation area and panelist pool.

Data collection and handling — This is another resource to be defined: the personnel, hardware, and software required to collect and treat the data to be generated by the panel. Topics such as the use of personal computers with PC software vs. the company's mainframe should be addressed before the data begin to accumulate on the sensory analyst's desk. The specific ways in which the data are generated and used (that is, frequency data, scalar data — category, linear, magnitude estimation), the number of attributes, the number of replications, and the need for statistical analysis, all contribute to the requirements for data collection and handling.

Projected costs — Once upper management and the project group understand the need to have a panel and the time and costs required for its development and use, the costs and benefits can be assessed from a business and investment perspective. This phase is essential so that the support from management is based on a full understanding of the panel development process. Once management and the project team are "on board", the sensory analyst can expect the support which is needed to satisfy the requirements for personnel, both panelists and staff, facilities, and data handling. Management can then, through circulars, letters, and/or seminars communicate its support for the development of and participation in sensory testing. As the reader will have gathered by now, public recognition by management of the importance of the sensory program and of the involvement of employees as panelists are essential for the operation of the system. If participation in sensory tests is not seen as a worthwhile expenditure of time by upper and middle management, the sensory analyst will find the recruiting task to be difficult if not impossible, and test participation will dry up more quickly than new recruits can be enrolled.

Once management support has been communicated through the organization and has been demonstrated in terms of facilities and personnel for the panel, the sensory analyst can use presentations, questionnaires, and personal contact to reach potential panel members. The time commitment and qualifications must be clearly iterated so that candidates understand what is required of them. General requirements include: interest in the test program, availability (about 80% of the time), promptness, general good health (no allergies or health problems affecting participation), articulateness, and absence of aversions to the product class involved. Other specific criteria are listed for individual tasks in Sections III and IV.

III. SELECTION AND TRAINING FOR DIFFERENCE TESTS

A. Selection

Assume that the early recruitment procedure has provided a group of candidates free of obvious drawbacks such as heavy travel or work schedules or health problems, which would make participation impossible or sporadic. The sensory analyst must now devise a set of screening tests which teach the candidates the test process while weeding out unsuitable nondiscriminators as early as possible. Such screening tests should *use the products to be studied* and *the sensory methods to be used in the study*. It follows that they should be

Table 1
SUGGESTED SAMPLES FOR MATCHING TESTS

Tastes, Chemical Feeling Factors

Flavor	Stimulus	Concentration (g/ℓ^a)
Sweet	Sucrose	20
Acid	Tartaric acid	0.5
Bitter	Caffeine	1.0
Salty	Sodium chloride	2.0
Astringent	Alum	10

Aroma, Fragrances, Odorants[b]

Aroma descriptors	Stimulus
Peppermint, minty	Peppermint oil
Anise, anethole, licorice	Anise oil
Almond, cherry, amaretto	Benzaldehyde, oil of bitter almond
Orange, orange peel	Orange oil
Floral	Linaloll
Ginger	Ginger oil
Jasmine	Jasmine-74-D-10%
Green	cis-3-Hexenol
Vanilla	Vanilla extract
Cinnamon	Cinnamaldehyde, Cassia oil
Clove, dentist's office	Eugenol, oil of Clove
Wintergreen, BenGay	Methyl salicylate, oil of Wintergreen

[a] In tasteless and odorless water at room temperature.
[b] Perfume blotters dipped in odorant, dried in hood 30 min, placed in wide mouth jar with tight cap.

patterned on those described below, rather than using them directly. The screening tests aim to determine differences among candidates in the ability (1) to discriminate (and describe, if attribute difference tests are to be used) character differences among products and (2) to discriminate (and describe with a scale for attribute difference tests) differences in the intensity or strength of the characteristic.

Suggested rules for evaluating the results are given at the end of each section. Bear in mind that while candidates with high success rates may on the whole be satisfactory, the best panel will result if selection can be based on potential rather than on current performance.

1. Matching Tests

Matching tests are used to determine a candidate's ability to discriminate (and describe, if asked in addition) differences among several stimuli presented at intensities well above threshold level. Familiarize candidates with an initial set of four to six coded but unidentified products. Then present a randomly numbered set of eight to ten samples, of which a subset is identical to the initial set. Ask candidates to identify on the scoresheet the familiar samples in the second set and to label them with the corresponding codes from the first set.

Table 1 contains a selection of samples suitable for matching tests. These may be common

Table 2
FRAGRANCE MATCHING TEST

Instructions: Sniff the first set of fragrances; allow time to rest after
each sample.
Sniff the second set of fragrances and determine which
samples in the second set correspond to each sample
in the first set.
Write down the code of the fragrance in the second
set next to its match from the first set.

Optional: (Determine which descriptor from the list below best
describes the fragrance pair.)

First set	Second set match	Descriptor[a]
079	_____	_____
318	_____	_____
992	_____	_____
467	_____	_____
134	_____	_____
723	_____	_____

[a]   A list of discriptors, similar to the one given below may be given
at the bottom of the scoresheet. The ability to select and use
descriptors should be determined if the candidates will be partic-
ipating in attribute difference tests.

Floral	Peppermint	Vanilla	
Green	Cinnamon	Ginger	
Jasmine	Orange	Cherry, Almond	Anise/Licorice

flavor substances in water, common fragrances, lotions with different fat/oil systems, products made with pigments of different colors, fabrics of similar composition but differing in
basis weight, etc. Care should be taken to avoid carryover effects, e.g., samples must not
be too strong. Table 2 shows an example of a scoresheet for matching fragrances at above
threshold levels in a nonodorous diluent.

2. Detection/Discrimination Tests

This type of selection test is used to determine a candidate's ability to detect differences
among similar products with ingredient or processing variables. Present candidates with a
series of three or more Triangle tests[5,6] with differences ranging from easy to moderately
difficult, see, e.g., Chapter 6, Section II.A, Example 3. Duo-trio tests (Chapter 6, Section
II.C) may also be used. Table 3 lists some common flavor standards and the levels at which
they may be used. "Doctored" samples, such as beers spiked[7] with substances imitating
common flavors and off-notes, may also be used. Arrange preliminary tests with experienced
tasters to determine the optimal order of the test series and to control that stimulus levels
are appropriate, detectable, but not overpowering. Use standard triangle or duo-trio scoresheets when suitable. If desired, use sequential triangle tests (Chapter 6, Section II.G) to
decide acceptance or rejection of candidates. However, as already mentioned, do not rely
too much on taste acuity.

3. Ranking/Rating Tests for Intensity

These tests are used to determine candidates' ability to discriminate graded levels of
intensity of a given attribute. Use rating on an appropriate scale if this is the method the

Table 3
SUGGESTED MATERIALS
FOR DETECTION TESTS

Substance	Concentration (g/ℓ[a])	
Caffeine	0.2[b]	0.4[c]
Tartaric acid	0.4[b]	0.8[c]
Sucrose	7.0[b]	14.0[c]
Δ-Decalactone	0.002[b]	0.004[c]

[a] Approximations of substances added
to tasteless and odorless water.
[b] 3 × and [c]6 × threshold levels.

Table 4
SUGGESTED MATERIALS FOR RANKING/RATING TESTS

	Sensory stimuli			
Taste				
Sour	Citric acid/water	0.25,	0.5,	1.0,	1.5 g/ℓ
Sweet	Sucrose/water	10,	20,	50,	100 g/ℓ
Bitter	Caffeine/water	0.3,	0.6,	1.3,	2.6 g/ℓ
Salty	Sodium chloride/water	1.0,	2.0,	5.0,	10 g/ℓ
Odor				
Alcoholic	3-Methylbutanol/water,	10,	30,	80,	180 mg/ℓ
Texture				
Hardness	Cream cheese,[a] American cheese,[a] peanuts, carrot slices[a]			
Fracturability	Corn muffin,[a] Graham cracker, Finn crisp bread, Life Saver			

[a] At ½″ thickness.

test panelist will eventually use; otherwise use ranking (Chapter 6, Section III.E). Present a series of samples in random order, in which one parameter is present at different levels, which cover the range present in the product(s) of interest. Ask candidates to rank the samples in ascending order (or rate them using the prescribed scale) according to the level of the stated attribute (sweetness, oiliness, stiffness, surface smoothness, etc.), see suggested materials in Table 4.

Typical scoresheets are shown in Tables 5 and 6. The selection sequence may make use of more than one attribute ranking/rating test, especially if the ultimate panel will need to cover several sense modalities, e.g., color, visual surface oiliness, stiffness, surface smoothness.

4. Interpretation of Results of Screening Tests

Matching tests — Reject candidates scoring less than 75% correct matches. Reject candidates for attribute tests who score less than 60% in choosing the correct descriptor.

Detection/discrimination tests — When using Triangle tests, reject candidates scoring less than 60% on the "easy" tests (6 × threshold) or less than 40% on the "moderately difficult" tests (3 × threshold). When using Duo-trio tests, reject candidates scoring less than 75% on the "easy" tests or less than 60% on the "moderately difficult" tests. Or use the sequential tests procedure as described in Chapter 6, Section II.G, Example 4.

Table 5
RANKING TEST FOR
INTENSITY

Rank the salty taste solutions in the coded cups in ascending order of saltiness.

Code

Least salty _____
_____
_____
Most salty _____

Table 6
RATING TEST FOR INTENSITY

Rate the saltiness of each coded solution for intensity/strength of saltiness using the line scale for each.

Code

463	None_____Strong
318	None_____Strong
941	None_____Strong
502	None_____Strong

Ranking/rating tests — Accept candidates ranking samples correctly or inverting only adjacent pairs. In the case of rating, use the same rank order criteria and expect candidates to use a large portion of the prescribed scale when the stimulus covers a wide range of intensity.

B. Training

To ensure development of a professional attitude to sensory analysis on the part of panelists, conduct the training in a controlled professional sensory facility. Instruct subjects how to precondition the sensory modality in question, e.g., not to use perfumed cosmetics and to avoid exposure to foods or fragrances for 30 min before sessions, how to prepare skin or hands for fabric and skin feel evaluations, and to notify the panel leader of allergic reactions which affect the test modality. On any day, excuse subjects suffering from colds, headaches, lack of sleep, etc.

From the outset, teach subjects the correct procedures for handling the samples before and during evaluation. Stress the importance of adhering to the prescribed test procedures, reading all instructions and following them scrupulously. Demonstrate ways to eliminate or reduce sensory adaptation, e.g., taking shallow sniffs of fragances and leaving several tens of seconds between sample evaluations. Stress the importance of disregarding personal preferences and concentrating on the detection of difference.

Begin by presenting samples of the product(s) understudy which represent large, easily perceived sensory differences. Concentrate initially on helping panelists to understand the scope of the project and to gain confidence. Repeat the test method using somewhat smaller but still easily perceived sample differences. Allow the panel to learn through repetition until full confidence is achieved.

For attribute difference tests, carefully introduce panelists to the attributes, the terminology used to describe them, and the scale method used to indicate intensity. Present a range of products showing representative intensity differences for each attribute.

Continue to train "on the job" by using the new panelist in regular discrimination tests. Occasionally introduce training samples to simulate "off notes" or other key product differences in order to keep the panel on track and attentive.

Be aware of changes in attitude or behavior on the part of one or more panelists who may be confused, losing interest, or distracted by problems at work or outside. The history of sensory testing is full of incredible results, which could have come only from panelists who were "lost" during the test with the sensory analyst failing to anticipate and detect a failure in the "test instrument".

IV. SELECTION AND TRAINING OF PANELISTS FOR DESCRIPTIVE TESTING

A. Selection for Descriptive Testing

When selecting panelists for descriptive analysis, the panel leader or panel trainer needs to determine each candidate's capabilities in three major areas:

1. For each of the sensory properties under investigation, such as fragrance or flavor or oral texture or skinfeel, the ability to detect differences in characteristics present and in their intensities
2. The ability to describe those characteristics using (a) verbal descriptors for the characteristics and (b) scaling methods for the differences in intensity
3. The capacity for abstract reasoning, as descriptive analysis depends heavily upon the use of references whose characteristics must be quickly recalled and applied to other products

In addition to screening panelists for the above descriptive capabilities, panel leaders must prescreen candidates for the following personal criteria:

1. Interest in full participation in the rigors of the training, practice, and ongoing work phases of a descriptive panel.
2. Availability to participate in 80% or more of all phases of the panel's work. Conflict with work load, travel, or even the candidate's supervisor may eventually cause the panelist to drop off the panel during or after training, thus losing one panelist from an already small number of 10 to 15.
3. General good health and no illnesses related to the sensory properties being measured, such as:
    a. Diabetes, hypoglycemia, hypertension, dentures, chronic colds or sinusitis, or food allergies in those candidates for flavor and/or texture analysis of foods, beverages, pharmaceuticals or other products for internal use.
    b. Chronic colds or sinusitis for aroma analysis of foods, fragrances, beverages, personal care products, pharmaceuticals, or household products.
    c. Central nervous system disorders or reduced nerve sensitivity due to the use of drugs affecting the central nervous system, for tactile analysis of personal care skin products, fabrics, or household products.

The ability to detect and describe differences, the ability to apply abstract concepts, and the degree of positive attitude and predilection for the tasks of descriptive analysis can all be determined through a series of tests which include

- A set of prescreening questionnaires
- A set of acuity tests
- A set of ranking/rating tests
- A personal interview

The investment in a descriptive panel is large in terms of time and human resources, and it is wise to conduct an exhaustive screening process, rather than training unqualified subjects.

Lists of screening criteria for three descriptive methods (the flavor profile, quantitative descriptive analysis, and texture profile) can be found in ASTM Special Technical Publication 758.[1] The criteria listed below are those used to select subjects for training in the Spectrum℠ method of descriptive analysis as described in Chapter 8. These can be applied to the screening of employees, or for external screening, in cases where recruiting from the local community is preferred because of time-consuming panels (20 to 50 hr per person per week). The additional prescreening questionnaires are used to select individuals who can verbalize and think in concepts. This reduces the risk of selecting outside panelists who have sensory acuity but cannot acquire the "technical" orientation of panels recruited from inside the company.

1. Prescreening Questionnaires

For a panel of 15, typically 40 to 50 candidates may be prescreened using questionnaires such as those shown here. Table 7 applies to a tactile panel (skinfeel or fabric feel); Table 8 to a food panel (flavor or texture), and Figure 1 tests the candidate's potential to learn scaling. Figure 1 can be used on its own, or in combination with Tables 7 or 8. Of the 40 to 50 original candidates, generally 20 to 30 qualify and proceed to the acuity tests.

2. Acuity Tests

To qualify for this stage, candidates should:

- Indicate no medical or pharmaceutical causes of limited perception
- Be available for the training sessions
- Answer 80% of the verbal questions correctly and clearly
- In the questionnaire Figure 1, assign scalar ratings which are within 10% of the correct value for all figures

Candidates should demonstrate ability to

- Detect and describe characteristics present in a qualitative sense
- Detect and describe intensity differences in a quantitative sense

Therefore, although detection tests (e.g., Triangle or Duo-trio tests using variations in formulation or processing of the product to be evaluated) may yield a group of subjects who can detect small product variables, detection alone is not enough for a descriptive panelist. To qualify, subjects must be able to adequately discriminate and describe some key sensory attributes within the modalities used with the product class under test and must show ability to use a rating scale correctly to describe differences in intensity.

Detection — The panel trainer presents a series of samples representing key variables within the product class, in the form of Triangle or Duo-trio tests.[6] Differences in process time or temperature (roast, bake, etc.), ingredient level (50%, 150% of normal), or packaging can be used as sample pairs to determine acuity in detection. Attempt to present the easier pairs of samples first and follow with pairs of increasing difficulty. Select subjects who achieve 50 to 60% correct replies in Triangle test, or 70 to 80% in Duo-trio tests, depending on the degree of difficulty of each test.

Description — Present a series of products showing distinct attribute characteristics (fragrance/flavor oils, geometrical texture properties[8,9]) and ask candidates to describe the sensory impression. Use the fragrance list in Table 1 without a list of descriptors from which to choose. The candidate must describe each fragrance using his/her own words. These may

Table 7

PRESCREENING QUESTIONNAIRE FOR A TACTILE PANEL (SKIN FEEL OR FABRIC FEEL)

PRESCREENING QUESTIONNAIRE
TACTILE: SKIN OR FABRIC FEEL

HISTORY

NAME: _____
ADDRESS: _____
PHONE, (HOME AND BUSINESS): _____

FROM WHAT GROUP OR ORGANIZATION DID YOU HEAR ABOUT THIS PROGRAM? _____

TIME:

1. ARE THERE ANY WEEKDAYS, (M—F), THAT YOU WILL NOT BE AVAILABLE ON A REGULAR BASIS? _____

2. HOW MANY WEEKS VACATION DO YOU PLAN TO TAKE BETWEEN JUNE 1—SEPTEMBER 30, 1985? _____

HEALTH:

1. DO YOU HAVE ANY OF THE FOLLOWING?
   CENTRAL NERVOUS SYSTEM DISORDER _____
   UNUSUALLY COLD OR WARM HANDS _____
   SKIN RASHES _____
   CALLOUSES ON HANDS/FINGERS _____
   HYPERSENSITIVE SKIN _____
   TINGLING IN THE FINGERS _____

2. DO YOU TAKE ANY MEDICATIONS WHICH AFFECT YOUR SENSES, ESPECIALLY TOUCH? _____

GENERAL:

1. IS YOUR SENSE OF TOUCH: WORSE THAN AVERAGE _____
   (CHECK ONE)   AVERAGE _____
           BETTER THAN AVERAGE _____

2. DOES ANYONE IN YOUR IMMEDIATE FAMILY WORK FOR A PAPER, FIBER OR TEXTILE COMPANY? _____
   A MARKETING RESEARCH OR ADVERTISING FIRM? _____

TACTILE/TOUCH QUIZ
ANSWER EACH QUESTION IN YOUR OWN WORDS, AS BEST YOU CAN

1. WHAT CHARACTERISTICS OF FEEL OF A TOWEL WOULD MAKE YOU THINK IT IS ABSORBANT? _____

2. WHAT IS THICKER, AN OILY OR GREASY FILM? _____

3. WHEN YOU RUB AN OILY FILM ON YOUR SKIN, HOW DO YOUR FINGERS MOVE?
   SLIP _____ OR DRAG _____ (CHECK ONE)

4. WHAT FEEL PROPERTIES WITH A TISSUE DO YOU ASSOCIATE WITH ITS SOFTNESS? _____

5. WHAT SPECIFIC APPEARANCE CHARACTERISTICS OF A BATH TISSUE INFLUENCE YOUR PERCEPTION OF THE FEEL OF IT? _____

6. NAME SOME THINGS THAT ARE STICKY: _____

7. WHEN YOUR SKIN FEELS MOIST, WHAT OTHER WORDS OR PROPERTIES COULD DESCRIBE IT? _____

8. NAME SOME THINGS THAT ARE ROUGH _____
   WHAT MAKES THEM ROUGH? _____

9. BRIEFLY, HOW WOULD YOU DEFINE "FULLNESS"? _____

10. WHAT DO YOU FEEL IN A FABRIC OR PAPER PRODUCT THAT MAKES IT FEEL STIFF? _____

11. WHAT OTHER WORDS WOULD YOU USE TO DESCRIBE A LOTION AS THIN OR THICK? _____

12. WHAT CHARACTERISTICS DO YOU FEEL WHEN YOU STROKE THE SURFACE OF A FABRIC? _____
    THE BACK OF YOUR HAND? _____

Table 8

PRESCREENING QUESTIONNAIRE FOR A FOOD/BEVERAGE PANEL

PRESCREENING QUESTIONNAIRE
FOODS/BEVERAGES

HISTORY

NAME: _____
ADDRESS: _____
PHONE, (HOME AND BUSINESS): _____

FROM WHAT GROUP OR ORGANIZATION DID YOU HEAR ABOUT THIS PROGRAM? _____

TIME:

1. ARE THERE ANY WEEKDAYS, (M–F), THAT YOU WILL NOT BE AVAILABLE ON A REGULAR BASIS? _____

2. HOW MANY WEEKS VACATION DO YOU PLAN TO TAKE BETWEEN JUNE 1–SEPTEMBER 1, 1985? _____

HEALTH:

1. DO YOU HAVE ANY OF THE FOLLOWING?
   DENTURES _____
   DIABETES _____
   ORAL DISEASE _____
   HYPOGLYCEMIA _____
   FOOD ALLERGIES _____
   HYPERTENSION _____

2. DO YOU TAKE ANY MEDICATIONS WHICH AFFECT YOUR SENSES, ESPECIALLY TASTE AND SMELL? _____

FOOD HABITS:

1. ARE YOU CURRENTLY ON A RESTRICTED DIET? IF YES, EXPLAIN. _____

2. HOW OFTEN DO YOU EAT OUT IN A MONTH? _____
3. HOW OFTEN DO YOU EAT FAST FOODS OUT IN A MONTH? _____
4. HOW OFTEN IN A MONTH, DO YOU EAT A COMPLETE FROZEN MEAL? _____
5. WHAT IS (ARE) YOUR FAVORITE FOODS(S)? _____

6. WHAT IS (ARE) YOUR LEAST FAVORITE FOOD(S)? _____

(FOOD HABITS – continued)

7. WHAT FOODS CAN YOU NOT EAT? _____

8. WHAT FOODS DO YOU NOT LIKE TO EAT? _____

9. IS YOUR ABILITY TO DISTINGUISH SMELL AND TASTES:

	SMELL	TASTE
BETTER THAN AVERAGE	___	___
AVERAGE	___	___
WORSE THAN AVERAGE	___	___

DOES ANYONE IN YOUR IMMEDIATE FAMILY WORK FOR A FOOD COMPANY? _____

DOES ANYONE IN YOUR IMMEDIATE FAMILY WORK FOR AN ADVERTISING OR MARKETING RESEARCH AGENCY? _____

FOOD QUIZ

1. IF A RECIPE CALLS FOR THYME AND THERE IS NONE AVAILABLE; WHAT WOULD YOU SUBSTITUTE? _____

2. WHAT ARE SOME OTHER FOODS THAT TASTE LIKE YOGURT? _____

3. WHY IS IT THAT PEOPLE OFTEN SUGGEST ADDING COFFEE TO GRAVY TO ENRICH IT? _____

4. HOW WOULD YOU DESCRIBE THE DIFFERENCE BETWEEN FLAVOR AND TEXTURE? _____

5. BRIEFLY, HOW WOULD YOU DESCRIBE THE DIFFERENCE BETWEEN CRISPY AND CRUNCHY? _____

6. WHAT IS THE BEST ONE OR TWO WORD DESCRIPTION OF GRATED ITALIAN CHEESE (PARMESAN OR ROMANO)? _____

7. DESCRIBE SOME OF THE NOTICEABLE FLAVORS IN MAYONNAISE. _____

8. DESCRIBE SOME OF THE NOTICEABLE FLAVORS IN COLA. _____

9. DESCRIBE SOME OF THE NOTICEABLE FLAVORS IN SAUSAGE. _____

10. DESCRIBE SOME OF THE NOTICEABLE FLAVORS IN RITZ CRACKERS. _____

include chemical terms (e.g., cinnamic aldehyde), common flavor terms (e.g., cinnamon), or related terms (e.g., like Red Hots candy, Big Red gum, Dentyne). Candidates should be able to describe 80% of the stimuli using chemical, common, or related terms and should at least attempt to describe the remainder with less specific terms (e.g., sweet, brown spice, hot spice).

3. Ranking/Rating Screening Tests for Descriptive Analysis

Having passed the prescreening tests and acuity tests, the candidate is ready for screening with the actual product class and/or sensory attribute for which the panel is being selected. Candidates should rank or rate a number of products on a selection of key attributes, using the technique of the future panel. These tests can be supplemented with a series of samples which demonstrate increasing intensity of certain attributes, such as tastes and odors (see Table 4), or oral texture properties (see Chapter 8, Appendix 2, Texture section. Scale no. 5 is suitable, containing hardness standards from cream cheese = 1.0 to hard candy = 14.5; also scale no. 7 which contains standards for fracturability from corn muffins at 1.0 to Life Savers candy at 14.7). A questionnaire such as Table 9 is suitable. For certain skinfeel and fabric feel properties, samples may need to be selected from among commercial products and laboratory prototypes, representing increasing intensity levels of selected attributes. Choose candidates who can rate all samples in the correct order for 80% of the attributes scaled. Allow for reversal of adjacent samples only and check that candidates use most of the scale for at least 50% of the attributes tested.

4. Personal Interview

Especially for descriptive panels, a personal interview is necessary in order to determine whether candidates are well suited to the group dynamics and analytical approach. Generally, candidates who have passed the prescreening questionnaire and all of the acuity tests are interviewed individually by the panel trainee or panel leader. The objective of the interview is to confirm the candidate's interest in the training and work phases of the panel including his/her availability with respect to workload, supervisor, and travel, and also communication skills and general personality. Candidates who express little interest in the sensory programs as a whole and in the descriptive panel in particular should be excused. Individuals with very hostile or very timid personalities may also be excluded, as they may detract from the needed positive input of each panelist.

B. Training for Descriptive Testing

The important aspect of any training sequence is to provide a structured framework for learning based on demonstrated facts and to allow the students, in this case panelists, to grow both in skills and confidence. Most descriptive panel training programs require between 40 and 120 hr of training. The amount of time needed depends on the complexity of the product (wine, beer, and coffee panels require far more time than those evaluating powdered drink mixes or breakfast cereals), on the number of attributes to be covered (a short-version descriptive technique for quality control or storage studies, Chapter 8, Section VI, requires fewer and simpler attributes), and on the requirements for validity and reliability (a more experienced panel will provide greater detail with greater reproducibility).

1. Terminology Development and Introduction to Scaling

The panel leader or panel trainer in conjunction with the project team must identify key product variables to be demonstrated to the panel during the initial stages of training. The project team should prepare a prototype or collect from commercially available samples an array of products as a frame of reference, which represents as many as possible of the attribute differences likely to be encountered in the product category. The panel is first

SCALING EXERCISES

INSTRUCTIONS: MARK ON THE LINE AT THE RIGHT TO INDICATE THE PROPORTION OF
THE AREA THAT IS SHADED.

EXAMPLES.

NONE|_____|ALL

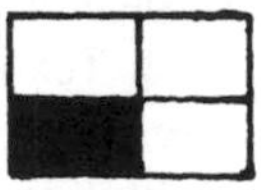

NONE|_____|ALL

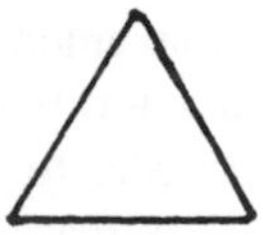

NONE|_____|ALL

1.    NONE|_____|ALL

2.    NONE|_____|ALL

3.    NONE|_____|ALL

4. 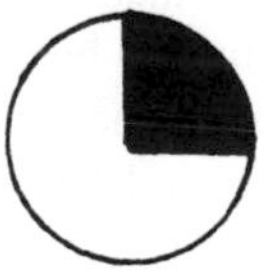   NONE|_____|ALL

5.   NONE_____ALL

6. 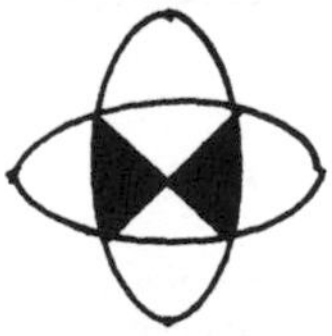   NONE_____ALL

7.    NONE_____ALL

8. 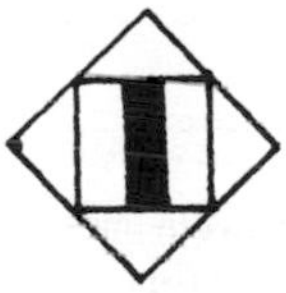   NONE_____ALL

9.    NONE_____ALL

10.    NONE_____ALL

FIGURE 1.   Prescreening questionnaire: scaling exercise. The answers are

1. 7/8	6. 1/4
2. 1/8	7. 3/4
3. 1/6	8. 1/8
4. 1/4	9. 2/3
5. 7/8	10. 1/2

Table 9
SCORESHEET CONTAINING TWO RANKING TESTS USED TO SCREEN CANDIDATES FOR A TEXTURE PANEL

Descriptive Texture Panel Screening

1. Place one piece of each product between *molars;* bite through *once;* evaluate for hardness. Rank the samples from least hard to most hard.

Least Hard   _____

_____

_____

_____

Most Hard   _____

2. Place one piece of each product between molars; bite down once and evaluate for crispness (crunchiness).

Least Crisp   _____

_____

_____

_____

Most Crisp   _____

introduced to the chemical (olfaction, taste, chemical feeling factors) and physical principles (rheological, geometrical, etc.) which govern or influence the perception of each product attribute. With these concepts and terms as a foundation, the panel then develops procedures for evaluation and terminology with definitions and references for the product class.[10,11]

This first stage of training may require up to 20 hr with the panel before a full list of descriptors is assembled. The scaling method may then be introduced, based on an array of samples which represent weak and strong intensities of some of the major attributes. In the case of flavor panels, the references listed in Chapter 8, Appendix 2 may be useful to demonstrate relative intensities across modalities (taste, olfaction/aromatics, and feeling factors).

2. Initial Practice

Once the panel has a grasp on the terminology and a general understanding of the use of each scale, the panel trainer or leader presents the panel with a series of samples (two or more) to be evaluated one at a time, which represent a *very* wide spread in qualitative (attributes) and quantitative (intensity) differences. At this early stage of development, which may be 15 to 40 hr, the panel must gain basic skills and confidence. The disparate samples allow the panel to see that the terms and the scale are effective as descriptors and discriminators and help the members to gain confidence both as individuals and as a group.

3. Small Product Differences

With the help of the project/product team, the panel leader collects samples which represent smaller differences within the product class, including variations in production variables and/or bench modifications of the product. The panel is encouraged to refine the procedures for evaluation and terminology with definitions and references to meet the needs of detecting and describing product differences. Care must be taken to reduce variations between supposedly identical samples; panelists in training tend to see variability in results as a reflection of their own lack of skill. Sample consistency contributes to panel confidence. This stage represents 10 to 15 hr of panel time.

4. Final Practice

The panel should continue to test and describe several products during the final practice stage of training (15 to 40 hr). The earlier samples should be fairly different and the final products tested should approach the real world testing situations for which the panel will be used.

During all four stages of the training program, panelists should meet after each session and discuss results, resolve problems or controversies, and ask for additional qualitative or quantitative references for review. This type of interaction is essential for developing the common terminology, procedures for evaluation, and scaling techniques which characterize a finely tuned sensory instrument.

V. PANEL PERFORMANCE AND MOTIVATION

Any good measuring tool needs to be checked regularly to determine its ability to perform validly and consistently. In the case of a sensory panel, the individuals need to be monitored, as well as the panel as a whole. Panels are comprised of human subjects, who have other jobs and responsibilities in addition to their participation in the sensory program, and it is necessary to find ways to maintain the panelists' interest and motivation over long periods of product testing.

A. Performance

For both difference and descriptive panels, the sensory analyst needs to have a measure of the performance of each panelist and of the panel, in terms of validity and reproducibility. Validity is the correctness of the response. In certain difference tests, such as the Triangle and Duo-trio, and in some directional attribute tests, the analyst knows the correct answer (the odd sample, the coded reference, the sweeter sample) and can assess the number of correct responses over time. The percent correct responses can be computed for each panelist on a regular monthly or bimonthly basis. Weighted scores can also be calculated, based on the difficulty of each test in which the panelist participated.[12] For the panel as a whole, validity can be measured by comparing panel results to other sensory test data, instrumental data, or the known variation in the stimulus, such as increased heat treatment, addition of a chemical, etc.

Reliability, or the ability to reproduce results, can be easily assessed for the individual panelists and for the panel as a whole by replicating the test, using duplicate test samples, or using blind controls.

For descriptive data, which are analyzed statistically by the analysis of variance, the panelists' performance can be assessed across each attribute as part of the data analysis, see ASTM STP 758,[1] pp. 29 to 30 for a detailed description of this analysis applied to a set of QDA results.

B. Feedback and Motivation

One of the major sources of motivation for panelists is a sense of doing meaningful work. After a project is completed, panelists should be informed by letter or a posted circular, of the project and test objectives, the test results, and the contribution made by the sensory results to the decision taken regarding the product. Immediate feedback after each test also tends to give the individual panelist a sense of "how am I doing?". The fears of some project leaders that panelists might become discouraged in tests with a low probability of success (a triangle test often has fewer than 50% correct responses) have proven groundless. Panelists do take into account the complexity of the sample, the difficulty of the test, and the probability of success. Panelists do want to know about the test and can indeed learn from past performance. Discussion of results after a descriptive panel session is highly recommended. The need to constantly refine the terms, procedures, and definitions is best served by regular panel interaction, after all the data have been collected.

C. Rewards and Motivation

In addition to the psychological rewards derived from feedback, panelists also respond positively and are further motivated to participate enthusiastically by a recognition and/or reward system. The presentation of certificates of achievement for:

- High panel attendance
- High panel performance
- Improved performance
- Completion of a training program
- Completion of a special project

stimulates panel performance and communicates to panelists that the evaluation is recognized as worthwhile. Short-term rewards, such as snacks, tokens for company products, and raffle tickets for larger prizes are often given daily to subjects. Over the longer term, sensory analysts often sponsor parties, outings, luncheons, or dinners for panelists, if possible with talks by project or company management describing how the results were used. Publicity for panel work in the company newspaper or the local community media serves to recognize the current panel members and stimulates inquiry from potential candidates. The underlying support by management for the full sensory program and for the active participation by panelists is a key factor in recruiting and maintaining an active pool of highly qualified members.

REFERENCES

1. ASTM, Committee E-18, *Guidelines for the Selection and Training of Sensory Panel Members*, ASTM Special Technical Publication 758, American Society for Testing and Materials, Philadelphia, 1981.
2. International Organization for Standardization, TC34/SC12, *Sensory Analysis — Choosing and Training Assessors*, Draft International Standard ISO/DP 8586, ISO, Tour Europe, Paris, 1985.
3. International Organization for Standardization, TC34/SC12, *Sensory Analysis — Determination of Sensitivity of Taste*, International Standard ISO 3972/1979, ISO, Tour Europe, Paris, 1979.
4. **Civille, G. V. and Carlton, D.,** Sensory Evaluation Panel Leadership Workshop, Morristown, N.J., 1985, 1986.
5. **Rainey, B.,** Selection and Training of Panelists for Sensory Panels, IFT Shortcourse: Sensory Evaluation Methods for the Practicing Food Technologist, St. Louis, Mo., Atlanta, Ga., Boston, Mass., and Portland, Ore., 1979.
6. **Zook, K. and Wessman, C.,** The selection and use of judges for descriptive panels, *Food Technol.*, 31(11), 56, 1977.
7. **Meilgaard, M. C., Reid, D. S., and Wyborski, K. A.,** Reference standards for beer flavor terminology system, *J. Am. Soc. Brew. Chem.*, 40, 119, 1982.
8. **Civille, G. V. and Szczesniak, A. S.,** Guidelines to training a texture profile panel, *J. Texture Stud.*, 6, 19, 1975.
9. **Szczesniak, A. S.,** Classification of textural characteristics, *J. Food Sci.*, 28, 385, 1963.
10. **Civille, G. V. and Liska, I. H.,** Modification and applications to foods of the General Foods sensory texture profile technique, *J Texture Stud.*, 6, 19, 1975.
11. **Schwartz, N.,** Adaptation of the sensory texture profile method to skin care products, *J. Texture Stud.*, 6, 33, 1975.
12. **Aust, L. B.,** Computers as an aid in discrimination testing, *Food Technol.*, 38(9), 71, 1984.

Chapter 11

BASIC PROBABILITY AND STATISTICAL METHODS

I. INTRODUCTION

The field of applied statistics can be divided into two general areas, estimation and inference. This chapter briefly presents the concepts and techniques of statistical estimation and inference as they relate to some of the more fundamental statistical methods used in sensory evaluation. The topics are presented with a minimum of theoretical detail. As such, those interested in pursuing this area further are encouraged to read any of the many good texts on applied statistics, particularly, Gacula and Singh,[1] Ott,[2] or, for a more theoretically advanced presentation, Snedecor and Cochran.[3]

Several definitions presented at this point will make the discussion that follows easier to understand. A statistical "population" is the entire collection of elements of interest. An "element" or "unit" from the population might be an individual, such as the consumer of a specific product, or a package of some product with a specific buy-before date. Measurements taken on elements from a population may be discrete, that is, take on only specific values, such as whole numbers, or continuous, that is, take on any value. The values that the measurements take on are goverened by a probability "distribution", usually expressed in the form of a mathematical equation, that relates the occurrence of a specific value to the probability of that occurrence. Associated with the distribution are certain fixed quantities called "parameters". The values of the parameters provide information about the population. In the normal distribution, for instance, the mean (μ) locates the center of the measurements. The standard deviation and variance (σ and σ^2, respectively) are measures of the dispersion or "spread" of the measurements about the mean. The "population proportion" (p) of a binomial distribution is the ratio of the number of elements in the population that possesses a characteristic (e.g., preference for some product) to the total number of elements in the population.

In the field of applied statistics, it is assumed that the form of the distribution is known but the values of the parameters (or at least some of them) are unknown. (This is what separates the field of statistics from its sister science probability. In the field of probability both the form of the distribution and the values of the parameters are known.) A subset of the elements of the population, called a "sample", is collected and the measurement or measurements of interest are made on each element in the sample. Mathematical functions of these measurements, called "statistics", are used to estimate the unknown values of the population parameters. The value of a statistic is called an "estimate".

Often a researcher is interested in determining if it is reasonable to believe that a specific situation, called the null hypothesis, exists in the population (e.g., two products, A and B, are equally preferred). Based on knowledge of the form of the probability distribution and on the initial assumption that the null hypothesis is true, the researcher uses the information contained in a sample (expressed in the form of statistics) to calculate the probability that the specified situation actually exists. If the calculated probability is very small the researcher concludes that the situation specified in the null hypothesis is not an accurate description of the state of the population. This process is called "statistical inference".

The remainder of the chapter is devoted to the further development of the ideas just presented. Section II deals with the basic concepts and techniques for calculating probabilities from some common distributions. Section III deals with the techniques of estimation. Section IV uses these techniques, and others, to present some concepts and methods for drawing statistical inferences about a population.

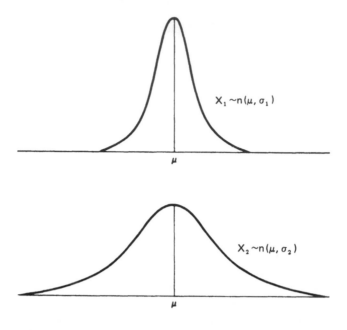

FIGURE 1.   Comparison of two normal distribution curves with the same mean but with $\sigma_1 < \sigma_2$.

II. PROBABILITY

The purpose of this section is to present the techniques for calculating probabilities based on some commonly used probability distributions. The techniques lay the foundation for statistical estimation and inference to be discussed later in this chapter, as well as for the more advanced topics discussed in Chapter 12.

A. The Normal Distribution

The "Normal Distribution" is among the most commonly used distributions in probability and statistics. The form of the normal distribution function is

$$f(X) = [1/\sqrt{2\pi}\sigma] \exp(-[X - \mu]^2/2\sigma^2)$$

where exp is the exponetial function with base e. The parameters of the normal distribution are the mean μ ($-\infty < \mu < \infty$) and the standard deviation σ ($\sigma > 0$). The normal distribution is symmetric about μ, that is, $f(x - \mu) = f(\mu - x)$. The mean μ measures the central location of the distribution. The standard deviation σ measures the dispersion or "spread" of the normal distribution about the mean. For small values of σ the graph of the distribution is narrow and peaked; for large values of σ the graph is wide and flat (see Figure 1). As with all continuous probability distributions, the total area under the curve is equal to one, regardless of the values of the parameters.

Let X be a random variable having a normal distribution with mean μ and standard deviation σ (often abbreviated as $X \sim n(\mu,\sigma)$). Define the variable Z as

$$Z = (X - \mu)/\sigma \tag{1}$$

The random variable Z also has a normal distribution. The mean of Z is zero and its standard deviation is one (i.e., $Z \sim n(0,1)$). Z is said to have a "standard normal distribution", or

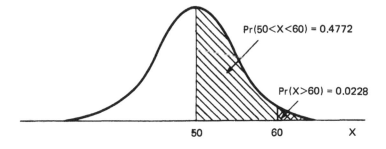

FIGURE 2. Calculating normal probabilities on an interval and in the tail of the distribution.

often times Z is called a "standard normal deviate". Given the values of μ and σ of a normal random variable X and a table of standard normal probabilities (Table T3*), it is possible to calculate various probabilities of interest.

1. Example 1: Normal Probabilities on an Interval

Consider the problem of calculating the probability of a normal random variable X where mean $\mu = 50$ and standard deviation $\sigma = 5$ take on a value between 50 and 60 (notationally, Pr[50<X<60]). The first step in solving the problem is to "standardize" X using Equation 1,

$$\text{Pr}[50 < X < 60] = \text{Pr}[(50 - 50)/5 < (X - 50)/5 < (60 - 50)/5]$$

$$= \text{Pr}[0 < Z < 2]$$

Table T3 gives the probabilities of a standard normal deviate taking on a value from zero (i.e., its mean) to some specified number. Therefore, entering Table T3 in the row corresponding to 2.0 and the column corresponding to 0.00 we find that the probability sought is equal to 0.4772 (see Figure 2).

Next consider the problem of finding Pr[45 < X < 50], where, as before X ~ n(50, 5). Standardizing, we find

$$\text{Pr}[45 < X < 50] = \text{Pr}[(45 - 50)/5 < (X - 50)/5 < (50 - 50)/5]$$

$$= \text{Pr}[-1 < Z < 0]$$

Because the standard normal distribution is symmetric about its mean, zero, it follows that Pr[$-c < Z < 0$] = Pr[0 < Z < c] for any constant c. Therefore, by Table T3,

$$\text{Pr}[-1 < Z < 0] = \text{Pr}[0 < Z < 1]$$

$$= 0.3413 \qquad (2)$$

So, we find that Pr[45 < X < 50] = 0.3413.

Finally, consider Pr[45 < X < 60] for the same random variable X ~ n(50, 5). This problem is solved as follows:

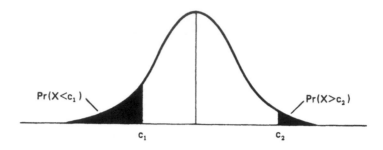

FIGURE 3.   Tail probabilities of a normal distribution.

$$\Pr[45 < X < 60] = \Pr[-1 < Z < 2] \quad \text{(Standardizing by Equation 1)}$$

$$= \Pr[-1 < Z < 0] + \Pr[0 < Z < 2]$$

$$= \Pr[0 < Z < 1] + \Pr[0 < Z < 2] \quad \text{(by Equation 2)}$$

$$= 0.3413 + 0.4772 \quad \text{(from Table T3)}$$

$$= 0.8185$$

Breaking the problem into two parts as is done in the second step, above, is necessary when the probability table being used states probabilities relative to the mean, zero (as does Table T3). Some probability tables, however, state probabilities relative to $-\infty$. The probability corresponding to $Z = 0.00$ will be 0.5000 in such tables.

2. Example 2: Normal Tail Probabilities

Tail probabilities are associated with the areas under the probability curve at the extremes of the distribution (see Figure 3). Notationally, tail probabilities are stated as $\Pr[X > c]$ or $\Pr[X < c]$ for some constant c. Tail probabilities are widely used in testing statistical hypotheses.

Consider the problem of finding $\Pr[X > 60]$, where $X \sim n(50, 5)$. Noting that the total area (i.e., probability) under any probability curve is one, it follows from the symmetry of the normal distribution that $\Pr[X < \mu] = \Pr[X > \mu] = 0.50$. Therefore,

$$\Pr[X > 60] = \Pr[(X - 50)/5 > (60 - 50)/5]$$

$$= \Pr[Z > 2] \quad \text{(by Equation 1)}$$

$$= 0.50 - \Pr[0 < Z \leqslant 2]$$

$$= 0.50 - 0.4772 \quad \text{(from Example 1)}$$

$$= 0.0228$$

(See the cross-hatched area in Figure 2 for an understanding of the third step, above.)

B. The Binomial Distribution

The ''Binomial Distribution'' function is

$$b[X] = \binom{n}{x}p^X(1 - p)^{n-X} \tag{3}$$

for $X = 0, 1, 2, ..., n$; $n > 0$ and an integer; and $0 \leq p \leq 1$. The parameters of the binomial distribution are n = the number of trials and p = the probability of "success" on any trial. The choice of what constitutes a "success" on each trial is arbitrary. For instance, in a two-sample preference test (A vs. B) preference for A could constitute a success or preference for B could constitute a success. Regardless, $X_i = 1$ for $i = 1, 2, ...,$ n if the result of the ith trial is a success; $X_i = 0$, otherwise. In Equation 3 $X = \sum_{i=1}^{n} X_i$ is the total number of successes in n trials. Table T2 is used to calculate binomial probabilities for values of $n = 1, 2, ..., 20$ and $p = 0.05$ to 0.50 in increments of 0.05.

1. Example 3: Calculating Exact Binomial Probabilities

Suppose $n = 16$ people participate in a Two-Out-of-Five difference test. The probability of correctly selecting the two odd samples from among the five is $p = 0.10$ if there is no perceivable difference among the samples. To find the probability that exactly two ($x = 2$) of the participants make the correct selections (i.e., exactly 2 successes in 16 trials), one enters the row of Table T2 corresponding to $n = 16$ and $x = 2$ and reads the value of the probability from the column corresponding to $p = 0.10$. In this case $b[2] = 0.2745$.

To find the probability that between two and six participants (inclusive) make the correct selections, one simply adds the probabilities associated with each possible outcome. That is

$$Pr[2 \leq X \leq 6] = \sum_{k=2}^{6} b[k]$$

$$= b[2] + b[3] + b[4] + b[5] + b[6]$$

$$= 0.2745 + 0.1423 + 0.0514 + 0.0137 + 0.0028$$

$$= 0.4847$$

There are two methods for calculating tail probabilities using Table T2. Consider the problem of calculating that probability that less than three particpants make the correct selections

$$Pr[X < 3] = P[X = 0 \quad \text{or} \quad X = 1 \quad \text{or} \quad X = 2]$$

$$= b[0] + b[1] + b[2]$$

$$= 0.1853 + 0.3294 + 0.2745$$

$$= 0.7892 \tag{4}$$

On the other hand consider calculating the probability that at least three participants make the correct selections (i.e., three or more), that is

$$Pr[X \geq 3] = \sum_{k=3}^{16} b[k] \tag{5}$$

For this situation the approach used in Equation 4 requires a great deal of addition that can be avoided if one makes the following observation. The probability of all possible outcomes is one. Therefore the probability sought is

$$\Pr[X \geqslant 3] = 1 - \Pr[X < 3]$$

$$= 1 - 0.7892 \qquad \text{(from Equation 4)}$$

$$= 0.2108 \tag{6}$$

Typically, it is easy to decide on which method (4) or (6) to use by looking at the probabilities in Table T2.

2. Example 4: The Normal Approximation to the Binomial

For values of n greater than those presented in Table T2 the approximate value of a binomial probability can be calculated using the normal distribution. To use the methods of Section II.A, one must know the values of μ and σ. For the number of successes these are

$$\mu = np$$

$$\sigma = \sqrt{np(1 - p)} \tag{7}$$

Let $n = 36$ and $p = \frac{1}{3}$ and consider the problem of calculating the probability of at least 16 successes. From Equation 7 one computes

$$\mu = (36)(1/3) = 12$$

$$\sigma = \sqrt{36(1/3)(1 - 1/3)}$$

$$= \sqrt{8} = 2.828$$

Therefore, using the methods of Example 2,

$$\Pr[X \geqslant 16] = \Pr[(X - 12)/2.828 \geqslant (16 - 12)/2.828]$$

$$= \Pr[Z \geqslant 1.41]$$

$$= 0.5 - \Pr[0 < Z < 1.41]$$

$$= 0.5 - 0.4207$$

$$= 0.0793$$

One can also use the normal approximation to the binomial to calculate probabilities associated with the proportion of successes. For this case

$$\mu = p$$

$$\sigma = \sqrt{p(1 - p)/n} \tag{8}$$

In most sensory evaluation tests the number of trials is large enough so that the normal approximation gives adequately accurate results. A rule of thumb often used is that the normal approximation to the binomial is sufficiently accurate if both $np > 5$ and $n(1 - p) > 5$, that is, for the normal approximation to be reasonably accurate, the sample size n should be sufficiently large so that one would expect to see at least five successes and at least five failures in the sample results.

III. ESTIMATION TECHNIQUES

Two classes of estimates are discussed in this section. They are the "point estimate" (i.e., a single number that estimates the value of a parameter) and a type of interval estimate called a "confidence interval" (i.e., a range of values that has a known probability of containing the true value of a parameter). The discussion includes statistics used to generate estimates of the parameters of two distributions that commonly occur in sensory evaluation, the normal distribution and the binomial distribution.

The parameters of the normal distribution are the mean μ and the standard deviation σ which is the positive square root of the variance σ^2. Methods for estimating both μ and σ are presented. The parameters of the binomial distribution are the number of trials n and the probability of "success" on any trial p.

A. Estimating the Parameters of a Normal Distribution

The arithmetic mean (or "sample mean") $\overline{X}$ (X-bar) is the statistic used to estimate the population mean of a normal distribution, where

$$\overline{X} = (\sum_{i=1}^{n} X_i)/n$$

$$= (X_1 + X_2 + ... + X_n)/n \tag{9}$$

Σ in Equation 9 is the sum function. The subscript (i = 1) and superscript (n) indicate the range over which the summing is to be done. Equation 9 indicates that the sum is taken over the measurements for all n elements in the sample.

The sample variance S^2, where

$$S^2 = (\sum_{i=1}^{n} X_i^2 - (\sum_{i=1}^{n} X_i)^2/n)/(n - 1) \tag{10}$$

is the statistic used to estimate σ^2, the variance of the population with a normal distribution. The positive square root of S^2, $S = \sqrt{S^2}$ is the statistic used to estimate the population standard deviation σ.

The sample mean $\overline{X}$ is a random variable. If the raw data used to calculate $\overline{X}$ are normal random variables, then $\overline{X}$ is also a normal random variable. In fact, even if the raw data are not distributed as normal random variables, $\overline{X}$ is still distributed approximately as a normal random variable providing the sample size used to calculate $\overline{X}$ is greater than 25 or so. Regardless, the mean of the distribution of $\overline{X}$ is the same as the mean of the distribution of the data used to calculate $\overline{X}$. Further, if σ is the standard deviation of the raw data's distribution, then $\sigma/\sqrt{n}$ is the standard deviation of the distribution of $\overline{X}$. The standard deviation of the distribution of $\overline{X}$ is called the "standard error of the mean". Notice that if the sample size n is increased then the standard error of the mean decreases. What this means in practice is that as the sample size becomes larger $\overline{X}$ is more and more likely to fall close to the true value of the population mean (see Figure 1). The standard error of the mean is estimated by $S_{\overline{x}} = S/\sqrt{n}$, where S is the positive square of S^2 in Equation 10.

1. Example 5: Estimating the Average Perceived Sweetness Intensity in a Cola and Its Standard Deviation

Suppose ten panelists rated the sweetness intensity of a cola sample using a 15-cm unstructured line scale. The calculations required to compute the sample mean and standard deviation of the sample values are presented in Table 1.

B. Estimating the Population Proportion p of a Binomial Distribution

The statistic used to estimate the population proportion p of a binomial distribution is $\hat{p}$ (p-hat), where

Table 1
EXAMPLE CALCULATIONS OF X̄ AND S

Panelist (i)	Sweetness (X$_i$)	(X$_i^2$)
1	10.4	108.16
2	10.7	114.49
3	10.0	100.00
4	10.2	104.04
5	10.3	106.09
6	10.3	106.09
7	10.6	112.36
8	10.4	108.16
9	10.5	110.25
10	10.3	106.09
Totals	$\sum_{i=1}^{10} X_i = 103.7$	$\sum_{i=1}^{10} X_i^2 = 1075.73$

$$\bar{X} = (\sum_{i=1}^{10} X_i)/n \qquad S^2 = (\sum_{i=1}^{10} X_i^2 - (\sum_{i=1}^{10} X_i)^2/n)/(n-1)$$

$$= 103.7/10 \qquad\qquad = (1075.73 - (103.7)^2/10)/9$$

$$= 10.37 \text{cm} \qquad\qquad = 0.04$$

$$S = \sqrt{S^2} = \sqrt{0.04} = 0.20 \text{ cm}$$

$$\hat{p} = (\text{number of ``successes''})/(\text{number of trials}) \qquad\qquad (11)$$

1. Example 6: A Preference Test

Suppose 200 consumers participate in a preference test between two samples, A and B. Further, suppose that 125 of the participants say that they prefer sample A. Preference for sample A was defined as a success before the test was conducted, so from Equation 11

$$\hat{p} = 125/200 = 0.625$$

That is, 0.625 or 62.5% is the estimated value of the proportion of consumers who prefer sample A.

C. Confidence Intervals on μ and p

Point estimates like those calculated above provide no information as to their own precision. Although they are a good starting point for the statistical analysis of a set of data it is always advisable to calculate confidence intervals on the unknown values of the parameters of interest. A confidence interval is a range of values within which the true value of a parameter lies with a known probability. Confidence intervals allow the researcher to determine if the calculated point estimates are sufficiently precise to meet the needs of an investigation.

Three types of confidence intervals are presented: the one-tailed upper confidence intervals, the one-tailed lower confidence interval, and the two-tailed confidence interval. The equations for calculating these intervals for both μ and p are presented in Table 2.

The quantities $t_{\alpha, n-1}$ and $t_{\alpha/2, n-1}$ in Table 2 are t-statistics. The quantity α measures the level of confidence. For instance, if α = 0.05 then the confidence interval is a 100 (1 − α) % = 95% confidence interval. The quantity (n − 1) in Table 2 is a parameter associated with the t-distribution called degrees of freedom. The value of t depends on the value of α and the number of degrees of freedom (n − 1). Critical values of t are presented in Table T4.

Table 2
COMPUTATIONAL FORMS FOR CONFIDENCE INTERVALS

Type of confidence interval	Parameter	
	μ	p
One-tailed upper	$\overline{X} + t_{\alpha,n-1}\sqrt{S^2/n}$	$\hat{p} + z_{\alpha}\sqrt{\hat{p}(1 - \hat{p})/n}$
One-tailed lower	$\overline{X} - t_{\alpha,n-1}\sqrt{S^2/n}$	$\hat{p} - z_{\alpha}\sqrt{\hat{p}(1 - \hat{p})/n}$
Two-tailed	$\overline{X} \pm t_{\alpha/2,n-1}\sqrt{S^2/n}$	$\hat{p} \pm z_{\alpha/2}\sqrt{\hat{p}(1 - \hat{p})/n}$

The quantity z in Table 2 is the critical value of a standard normal variable. The standard normal distribution has mean $\mu = 0$ and standard deviation $\sigma = 1$. Critical values for z for some commonly used levels of α are presented in the last row of Table T4 (i.e., the row corresponding to ∞ degrees of freedom).

1. Example 7: Confidence Intervals on the Mean μ

Consider the sweetness intensity data presented in Table 1. The sample mean intensity was $\overline{x} = 10.37$ cm and the sample standard deviation of the data was s = 0.20 cm. To construct a lower one-tailed 95% confidence interval on the value of the population mean one uses Table 2 and Table T4 to obtain

$$\overline{X} - t_{\alpha,n-1}\sqrt{s^2/n}$$

where $\alpha = 0.05$ and n = 10, thus yielding

$$10.37 - 1.883\sqrt{0.04/10}$$

$$= 10.37 - 0.12$$

$$= 10.25$$

The limit is interpreted to mean that the researcher is 95% sure that the true value of the population mean sweetness intensity is no less than 10.25 cm.

A two-tailed 95% confidence interval on the mean is calculated as

$$\overline{x} \pm t_{\alpha/2,n-1}\sqrt{s^2/n}$$

where n = 10 and $\alpha = 0.05$, so $t_{\alpha/2, n-1}$ is $t_{0.025, 9} = 2.262$, yielding

$$10.37 \pm 2.262(0.0633)$$

$$= 10.37 \pm 0.14$$

or

$$(10.23, 10.51)$$

That is, the researcher is 95% sure that the true value of the mean sweetness intensity lies on the interval from 10.23 cm to 10.51 cm.

2. Example 8: Confidence Interval on the Proportion p

Consider the consumer preference test in Example 6 where 125 of the 200 ($\hat{p} = 0.625$)

consumers preferred sample A. To construct a 95% confidence interval (two-tailed) on the true value of the population proportion p one uses Table 2 and Table T4 to obtain

$$\hat{p} \pm z_{\alpha/2}\sqrt{\hat{p}(1 - \hat{p})/n}$$

where n = 200, α = 0.05, and $\hat{p}$ = 0.625, yielding

$$0.625 \pm 1.96\sqrt{(0.625)(0.375)/200}$$

or

$$(0.558, 0.692)$$

The researcher may conclude, with 95% confidence, that the true proportion of the population that prefers sample A lies between 55.8% and 69.2%.

IV. STATISTICAL INFERENCE

Often the objective of an investigation is to determine if it is reasonable to assume that the unknown value of a parameter is equal to some specified value or possibly that the unknown values of two parameters are equal to each other. Determinations of this sort are made using the techniques of statistical inference. The type of inference discussed in this section is called hypothesis testing.

The process of statistical hypothesis testing is summarized by the following five steps:

1. The objective of the investigation is stated in mathematical terms, called the "null hypothesis" (H₀) (e.g., H_0: μ = c, for some constant, c).
2. Based on the prior interest of the researcher, another mathematical statement, called the alternative hypothesis (H_a) is formulated (e.g., H_a: μ>c, H_a: μ<c, or H_a: $\mu \neq$c).
3. A sample of elements from the population is collected and the measurement of interest is taken on each element of the sample.
4. The value of the statistic used to estimate the parameter of interest is calculated.
5. Based on knowledge of the probability distribution of the measurements and the assumed (null hypothesis) value of the parameter of interest, the probability that the statistic takes on the value calculated in step 4 is computed. If this probability is smaller than some predetermined value (α), the null hypothesis is rejected in favor of the alternative hypothesis.

A. Statistical Hypotheses

As stated above, statistical hypotheses are mathematical statements that typically specify the value of some parameter in a probability distribution, such as the mean μ of a normal distribution or the population proportion p of a binomial distribution. The null hypothesis is determined by the objective of the investigation. The value or values specified in the null hypothesis is used to calculate the probability used in the hypothesis test. The alternative hypothesis is developed based on the prior interest of the investigator. For example, if a company is replacing one of the raw ingredients in its current product with a less expensive ingredient from an alternate supplier, the sensory analyst's only interest going into the study would be to determine with a high level of confidence that the product made with the less expensive ingredient is not less preferred than the company's current product. The null hypothesis and the alternative hypothesis for this investigation are

$$H_o: p_{current} = p_{less\ expensive}$$

vs.

$$H_a: p_{current} > p_{less\ expensive}$$

where p_i is the proportion of the population that prefers product i.

Both the null and the alternative hypotheses must be specified before the test is conducted. The results of the statistical tests are biased in favor of rejecting the null hypothesis more often than it should be if the alternative hypothesis is formulated after reviewing the data.

B. One-Sided and Two-Sided Hypotheses

There are two types of alternative hypotheses, one-sided alternatives and two-sided alternatives. Some examples of situations leading to one-sided and two-sided alternatives are

One-sided	Two-sided
Confirm that a test brew is more bitter	Decide which test brew is more bitter
Confirm that a test product is preferred (as we had reason to expect)	Decide which test product is preferred
In general, whenever the alternative hypothesis has the form: A is more (less) than B, where A and B are specified	In general, whenever the alternative hypothesis has the form: A is different from B

Researchers often have trouble deciding whether the alternative hypothesis is one-sided or two-sided. Rules of thumb that work for one person may misguide others. The general criteria presented here will hopefully make this decision an easier one.

There are no statistical criteria for deciding if an alternative hypothesis should be one-sided or two-sided. In general, the form of the alternative hypothesis is determined by the prior interest of the researcher. If the researcher is only interested in determining if two samples are different then the alternative hypothesis is two-sided. If, on the other hand, the researcher wants to test for a specific difference between two samples, that is, one sample (specified) is more preferred, or more sweet, etc. than another sample, then the alternative hypothesis is one-sided. Most alternatives are two-sided, unless the researcher states that a specific type of difference is of interest before the study is conducted.

Confusion as to whether the alternative hypothesis is one-sided or two-sided may arise as a result of translating the researcher's stated interests into the mathematical form of the alternative hypothesis. Consider the case of a researcher who wants to determine if there is an overall difference between two samples. On its face value this is a two-sided alternative. However, if the sensory analyst chooses the Triangle test method then the mathematical form of the alternative hypothesis is one-sided. $H_a: p > 1/3$. Only when the proportion of correct answers is significantly greater than 1/3 does the analyst conclude that the panelists are not guessing, but rather that a perceivable difference does in fact exist. However, the conclusion drawn from this study would be two-sided, that is, the conclusion would be that the samples are different, with no directional indications given. A subtle distinction needs to be made here to get us out of this apparent dilemma. The alternative hypothesis above is, in fact, two-sided. However, the statistical method used to test the hypothesis is one-tailed. In a Triangle test the null hypothesis is rejected only when the number of correct answers falls far into the upper tail of the binomial distribution. Similar situations arise when χ^2 and F tests are performed. The alternative hypothesis may be two-sided but the statistical method only rejects the null hypothesis for large values of the test statistic (i.e., one-tailed results).

In practice, researchers should express their interests (i.e., the null and alternative hy-

Table 3
TYPE I AND TYPE II ERRORS OF SIZE α
AND β

	Conclusion Drawn	
	H₀ rejected	**H₀ accepted**
H_0 true	Type I error	Correct decision
	Pr[Type I error] = α	
H_0 false	Correct decision	Type II error
		Pr[Type II error] = β

potheses) in their own words. If the researcher's interests are clearly stated it is normally easy to decide whether the alternative hypothesis is one-sided or two-sided. The sensory analyst should report the results of the study in terms of the researcher's stated interests (one-sided or two-sided) regardless of whether the statistical method is one-tailed or two-tailed.

C. Type I and Type II Errors

In testing statistical hypotheses some conclusion is drawn. The conclusion may be correct or incorrect. There are two ways in which an incorrect conclusion may be drawn. First, a researcher may conclude that the null hypothesis is false when, in fact, it is true. Such an error is called a Type I error. Second, a researcher may conclude that the null hypothesis is true, or more correctly that the null hypothesis cannot be rejected, when, in fact, it is false. Such an error is called a Type II error (see Table 3). Gardener has presented figures similar to Table 3 in which the practical implications of Type I and Type II errors are given for several common testing situations. Gardener's figures are reproduced in Figure 4a and 4b.

The probability of making Type I and Type II errors is specified before the investigation is conducted. These probabilities are used to determine the required sample size for the study (see, for example, Snedecor and Cochran,[3] p. 102). The probability of making a Type I error is equal to α. The probability of making a Type II error is equal to β. Although α and β are probabilities (i.e., numbers) it is currently a common practice to use Type I error and α-error (as well as Type II error and β-error) interchangeably. This somewhat casual use of terminology causes little confusion in practice.

D. Examples: Tests on Means, Standard Deviations, and Proportions

This section presents procedures for conducting routine tests of hypotheses on means and standard deviations of normal distributions and on the population proportion (or probability of success) p from binomial distributions.

1. Example 9: Testing that the Mean of a Normal Distribution is Equal to a Specified Value

Suppose in the sweetness example in Section III.A that the sensory analyst wanted to test whether the average sweetness of the sample was ten or greater than ten. The mathematical forms of the null hypothesis and alternative hypothesis are

$$H_o: \mu = 10$$

vs.

$$H_a: \mu > 10$$

The alternative hypothesis is one-sided.

a) In Difference Tests

	Decision	
Truth	Reject H_0	Do Not Reject H_0
H_0 TRUE	**TYPE I ERROR** - Substitution Takes Place When It Should Not - New Product Promotion Done On Same Product As Before - Franchise In Trouble Due To Loss Of Consumer Confidence	Correct Decision
H_0 FALSE	Correct Decision	**TYPE II ERROR** - Substitution Does Not Take Place When It Should - Candidate Sample Missed - Money, Effort, Time Is Lost - We "Missed The Boat"

b) In Similarity Tests

	Decision	
Truth	Reject H_0	Do Not Reject H_0
H_0 TRUE	**TYPE I ERROR** - Substitution Does Not Take Place When It Should - Candidate Sample Missed - Money, Effort, Time Is Lost - We "Missed The Boat"	Correct Decision
H_0 FALSE	Correct Decision	**TYPE II ERROR** - Substitution Takes Place When It Should Not - Franchise In Trouble Due To "Less Preferred" or "Unfamiliar" Product Being Introduced

FIGURE 4. Practical implications of Type I and Type II Errors according to Gardner in (a) Difference tests and in (b) Similarity tests.

The statistical procedure used to test this hypothesis is a one-tailed, one-sample t-test.* The form of the test statistic is

$$t = (\overline{X} - \mu_{H_0})/(S/\sqrt{n}) \qquad (12)$$

The values of $\overline{X}$ and S are calculated in Table 1. Substituting in Equation 12 yields

$$t = (10.37 - 10)/(0.20/\sqrt{10}) = 5.85$$

The value of t, above, is compared to the upper-α critical value of a t distribution with $(n - 1)$ degrees of freedom (denoted as $t_{\alpha, n - 1}$). The value of $t_{\alpha, n - 1}$ marks the point in the t distribution (with $(n - 1)$ degrees of freedom) for which the probability of observing any larger value of t is α. Suppose the sensory analyst decides to control the Type I error at 5% (i.e., $\alpha = 0.05$). Then, from the row of Table T4 corresponding to 9 degrees of freedom the value of $t_{0.05, 9} = 1.833$, so the sensory analyst rejects the null hypothesis assumption that $\mu = 10$ in favor of the alternative hypothesis that $\mu > 10$ at the 5% significance level.

If the alternative hypothesis H_a, above, had been H_a: $\mu \neq 10$ (i.e., a two-sided alternative), then the null hypothesis would be rejected for absolute values of t (in Equation 12) greater than $t_{\alpha/2, (n - 1)}$, that is, reject if $|t| > t_{0.025, 9} = 2.262$ (from Table T4).

2. Example 10: Comparing the Means of Two Normal Populations — Paired-Sample Case

Sensory analysts often compare two samples by having a single panel evaluate both samples. When each member of the panel evaluates both samples the paired t-test is the appropriate statistical method to use. In general the null hypothesis can specify any difference of interest (i.e., H_0: $\delta = \mu_1 - \mu_2 = \delta_0$). (Setting $\delta_0 = 0$ is equivalent to testing H_0: $\mu_1 = \mu_2$). The alternative hypothesis can be two-sided (i.e., H_a: $\delta = \mu_1 - \mu_2 \neq \delta_0$) or one-sided ($H_a$: $\delta > \delta_0$ or H_a: $\delta < \delta_0$). In any case, the form of the paired-t statistic is

$$t = \frac{\overline{\delta} - \delta_o}{S_\delta/\sqrt{n}} \qquad (13)$$

where $\overline{\delta}$ is the average of the differences between the two samples and S_δ is the sample standard deviation of the differences. Consider the data in Table 4, which summarize the scores of the panel on a single attribute. The analyst wants to test whether the average rating of sample 1 is more than two units greater than the average rating for sample 2. The null hypothesis in this case is H_0: $\delta \leq 2$ vs. the alternative hypothesis H_a: $\delta > 2$. The test statistic is calculated as

$$t = \frac{2.54 - 2.00}{0.61/\sqrt{10}} = 2.79$$

($n = 10$ is used as the sample size because there are ten judges, each contributing one difference to the data set.) The null hypothesis is rejected if the value of t, above, exceeds

---

* Tests on means can be performed using standard normal z values if the standard deviation of the normal population, σ, is known. t-tests are based on student's t distribution which is very similar to the normal distribution. The t distribution is used when σ is estimated by S because the t distribution takes into account that the estimated value of S may deviate slightly from σ.

Table 4
DATA AND SUMMARY STATISTICS FOR THE PAIRED *t*-TEST EXAMPLE 10

Judge	Sample 1	Sample 2	Difference
1	7.3	5.7	1.6
2	8.4	5.2	3.2
3	8.7	5.9	2.8
4	7.6	5.3	2.3
5	8.0	6.1	1.9
6	7.1	4.3	2.8
7	8.0	5.7	2.3
8	7.5	3.8	3.7
9	6.9	4.5	2.4
10	7.4	5.0	2.4

$$\bar{\delta} = 2.54$$
$$S_\delta = 0.61$$

the upper-α critical value of the *t*-distribution with $(n - 1)$ degrees of freedom (i.e., $t_{\alpha, n-1}$).

The analyst decides to set $\alpha = 0.05$ and finds in Table T4 that $t_{0.05, 9} = 1.833$. The value of $t = 2.79$ is greater than 1.833, so the analyst rejects the null hypothesis and concludes at the 5% significance level that the average rating for sample 1 is more than two units greater than the average rating for sample 2.

3. Example 11: Comparing the Means of Two Normal Populations — Independent (or Two-Sample) Case

Suppose that a sensory analyst has trained two descriptive panels at different times and that the analyst now wants to merge the two groups. The analyst wants a high level of confidence that the two groups score samples with equivalent ratings before merging the groups and treating them as one panel.

The sensory analyst conducts several attribute panels to ensure that the two groups are similar. For each attribute considered the analyst presents samples of the same product to all panelists and records their scores and the group they belong to. The data from one of the studies are presented in Table 5. The null hypothesis for this test is H_0: $\mu_1 = \mu_2$ (or, equivalently H_0: $\mu_1 - \mu_2 = 0$). The alternative hypothesis is H_a: $\mu_1 \neq \mu_2$ (i.e., two-sided alternative).

The test statistic used to test the hypothesis is a two-sample *t*-test. The form of the test statistic is

$$t = \frac{(\bar{X}_1 - \bar{X}_2) - \delta_o}{\sqrt{\dfrac{(n_1 - 1) S_1^2 + (n_2 - 1) S_2^2}{n_1 + n_2 - 2}} \sqrt{\dfrac{1}{n_1} + \dfrac{1}{n_2}}} \tag{14}$$

where δ_o is the difference specified in the null hypothesis ($\delta_o = 0$ in the present example). Substituting the values from Table 5 into Equation 14 yields

Table 5

**DATA AND SUMMARY
STATISTICS FOR THE TWO-
SAMPLE *t*-TEST EXAMPLE 11**

Group 1		Group 2	
Judge	Score	Judge	Score
1	6.2	1	6.7
2	7.5	2	7.6
3	5.9	3	6.3
4	6.8	4	7.2
5	6.5	5	6.7
6	6.0	6	6.5
7	7.0	7	7.0
		8	6.9
		9	6.1

$$n_1 = 7 \qquad n_2 = 9$$
$$\overline{X}_1 = 6.557 \qquad \overline{X}_2 = 6.778$$
$$S_1 = 0.580 \qquad S_2 = 0.460$$

$$t = \frac{(6.557 - 6.778) - 0}{\sqrt{\dfrac{(7 - 1)(0.580)^2 + (9 - 1)(0.460)^2}{7 + 9 - 2}} \sqrt{\dfrac{1}{7} + \dfrac{1}{9}}}$$

$$= \frac{-0.221}{\sqrt{0.265}\sqrt{0.254}} = -0.85$$

The value of $t = -0.85$ is compared to the critical value of a t distribution at the $\alpha/2$ significance level (because the alternative hypothesis is two-sided and the t-test procedure is two-tailed) with $(n_1 + n_2 - 2)$ degrees of freedom. For the present example (using $\alpha = 0.05$) $t_{0.025, 14} = 2.145$ from Table T4. The null hypothesis is rejected if the absolute value (i.e., disregard the sign) of t is greater than 2.145. Since the absolute value of $t = -0.85$ (i.e., $|t| = 0.85$) is less than $t_{0.025, 14} = 2.145$ the sensory analyst does not reject the null hypothesis and therefore assumes that the two groups report similar ratings for this attribute.

4. Example 12: Comparing Standard Deviations from Two Normal Populations

The sensory analyst in Example 11 should also be concerned that the variabilities of the scores of the two groups are the same. To test that the variabilities of the two groups are equal the analyst compares the standard deviations of the two groups. The null hypothesis for this test is $H_0: \sigma_1 = \sigma_2$. The alternative hypothesis is $H_a: \sigma_1 \neq \sigma_2$ (i.e., a two-sided alternative). The test statistic used to test this hypothesis is

$$F = \frac{S^2_{(Big)}}{S^2_{(Small)}} \tag{15}$$

where $S^2_{(Big)}$ is the square of the larger of the two sample standard deviations and $S^2_{(Small)}$ is the square of the smaller sample standard deviation. In Table 5 Group 1 has the larger sample standard deviation, so $S^2_{(Big)} = S^2_1$ and $S^2_{(Small)} = S^2_2$ for this example. The value of F in Equation 15 is then,

$$F = (0.58)^2/(0.46)^2 = 1.59$$

The value of F is compared to the upper $\alpha/2$ critical value of an F distribution with $(n_1 - 1)$ and $(n_2 - 1)$ degrees of freedom (the numerator degrees of freedom are $(n_1 - 1)$ because $S^2_{(Big)} = S^2_1$ for this example. If $S^2_{(Big)} = S^2_2$ then the degrees of freedom would be $(n_2 - 1)$ and $(n_1 - 1)$). Using a significance level of $\alpha = 0.05$ the value of $F_{0.025, 6, 8}$ is found in Table T6 to be 4.65. The null hypothesis is rejected if $F > F_{\alpha/2, (n_1 - 1), (n_2 - 1)}$. Since $F = 1.59 < F_{0.025, 6, 8} = 4.65$, the null hypothesis is not rejected at the 5% significance level. The sensory analyst concludes that there is not sufficient reason to believe the two groups differ in the variability of their scoring on this attribute.

This is another example of a two-sided alternative that is tested using a one-tailed statistical method. The criterion for two-sided alternatives is to reject the null hypothesis if the value of F in Equation 15 exceeds $F_{\alpha/2, v1, v2}$, where v_1 and v_2 are the numerator and denominator degrees of freedom, respectively. Equation 15 is still used for one-sided alternatives i.e., $H_a\sigma_1 > \sigma_2$, but the criterion becomes, "reject the null hypothesis if $F > F_{\alpha, v1, v2}$."

5. Example 13: Testing that the Population Proportion is Equal to a Specified Value

Suppose that two samples (A and B) are compared in a preference test. The objective of the test is to determine if either sample is preferred by more than 50% of the population. The sensory analyst collects a random sample of n = 200 people, presents the 2 samples to each person in a balanced, random order, and asks each person which sample they prefer.* "No-preference" responses are divided equally among the two samples and it is found that 125 of the people said they preferred Sample A. The estimated proportion of the population that prefer Sample A is then $\hat{p}_A = 125/200 = 62.5\%$ by Equation 11.

The sensory analyst arbitrarily picks "preference for Sample A" as a "success" and tests the hypothesis $H_0: \hat{p}_A = 50\%$ vs. the alternative $H_a: \hat{p}_A \neq 50\%$. The analyst chooses to test this hypothesis at the $\alpha = 0.01$ significance level, using the appropriate z test, of the form

$$z = \frac{\hat{p} - p_o}{\sqrt{(p_o)(1 - p_o)/n}} \quad \text{for } \hat{p} \text{ and } p_o \text{ proportions}$$

or

$$z = \frac{\hat{p} - p_o}{\sqrt{(p_o)(100 - p_o)/n}} \quad \text{for } \hat{p} \text{ and } p_o \text{ percentages} \tag{16}$$

where $\hat{p}$ and p_0 are the observed and hypothesized values of p, respectively. Substituting the observed and hypothesized values into Equation 16 yields

$$z = (62.5 - 50.0)/\sqrt{(50)(100 - 50)/200} = 3.54$$

The value of z, above, is compared to the critical value of a standard normal distribution. For two-sided alternatives the absolute value of z is compared to $z_{\alpha/2} = t_{\alpha/2, \infty}$, (for one-sided alternatives, the value of z is compared to $z_\alpha = t_{\alpha, \infty}$) using Table T4. The value of $z_{0.005} = t_{0.005, \infty} = 2.576$. Since $z = 3.54$ is greater than 2.576, the null hypothesis is rejected and the analyst concluded at the 1% significance level that Sample A is preferred by more than 50% of the population.

---

* Note that this example is quite similar to the paired *t*-test in Example 2, but in this case the data are ranks (i.e., either 0 or 1).

Table 6
RESULTS OF A TWO REGION PREFERENCE TEST

	Preference		
Region	Product A	Product B	Total
1	125	75	200
2	102	98	200
Total	227	173	400

6. Example 14: Comparing Two Population Proportions

Let us extend Example 5 to take into account regional preferences. Suppose a company wishes to introduce a new product (A) into two regions and wants to know if their product is equally preferred over their prime competitor's product (B) in both regions. The sensory analyst conducts a 200-person preference test in each region and obtains the results shown in Table 6.

The null hypothesis in this example is H_0: $p_1 = p_2$ vs. the alternative hypothesis H_a:$p_1 \neq p_2$, where p_i is defined as the proportion of the population in region i that prefers Product A. This hypothesis is tested using a χ^2 test of the form

$$\chi^2 = \sum_{i=1}^{r} \sum_{j=1}^{c} (O_{ij} - E_{ij})^2/E_{ij} \tag{17}$$

where r and c are the numbers of rows and columns in a data table like Table 6. O_{ij} is the observed value in row i and column j of a data table. E_{ij} is the "expected" value for the entry in the ith row and j^{th} column of the data table. The E_{ij} are calculated as

$$E_{ij} = (\text{Total for row i})(\text{Total for column j})/(\text{Grand total})$$

Substituting the values from Table 6 into Equation 17 we obtain

$$\chi^2 = \frac{(125 - (200)(227)/400)^2}{(200)(227)/400} + \frac{(75 - (200)(173)/400)^2}{(200)(173)/400}$$

$$+ \frac{(102 - (200)(227)/400)^2}{(200)(227)/400} + \frac{(98 - (200)(173)/400)^2}{(200)(173)/400}$$

$$= \frac{(125 - 113.5)^2}{113.5} + \frac{(75 - 86.5)^2}{86.5} + \frac{(102 - 113.5)^2}{113.5} + \frac{(98 - 86.5)^2}{86.5}$$

$$= 5.39 \tag{18}$$

The value of χ^2 in Equation 18 is compared to the upper-α critical value of a χ^2-distribution with $(r - 1)(c - 1)$ degrees of freedom. If the analyst chooses $\alpha = 0.10$ (i.e., 10% significance level) then the critical value of $\chi^2_{0.10, 1} = 2.71$. Since $\chi^2 = 5.39 > \chi^2_{0.10, 1} = 2.71$, the analyst concludes at the 10% significance level that product A is not equally preferred over product B in both regions. Regional formulations may have to be considered.

V. CONCLUDING REMARKS

The basic techniques of probability and statistics were presented. Using probability and statistical estimation the methods for testing simple hypotheses were presented along with examples of some common hypothesis testing situations.

Techniques like the ones discussed in this chapter are quite useful. However, the objectives of a study should never be modified to suit a preexisting test method. If none of the methods presented in this chapter are appropriate for the study being undertaken, then possibly one of the more advanced techniques discussed in Chapter 12 is what is needed.

REFERENCES

1. **Gacula, M. C. and Singh, J.,** *Statistical Methods in Food and Consumer Research,* Academic Press, Orlando, Fla., 1984.
2. **Ott, L.,** *An Introduction to Statistical Methods and Data Analysis,* Duxbury Press, Belmont, Calif., 1977.
3. **Snedecor, G. W. and Cochran, W. G.,** *Statistical Methods,* Iowa State University Press, Ames, 1980.

Chapter 12

ADVANCED STATISTICAL TECHNIQUES

I. INTRODUCTION

The basic statistical techniques of estimation and inference presented in Chapter 11 are all that would be required to analyze the results of many sensory tests. However, sensory analysts are often called upon to answer quite complicated research questions, questions that require very sophisticated statistical techniques to answer. This chapter presents some of the most common of these advanced techniques. The computational complexity of the techniques to be discussed makes hand calculation impractical. Therefore, it is assumed that the reader has access to computer resources capable of performing the necessary calculations. Computational forms of equations are not presented. Such forms are available from other sources (e.g., Kirk[1]).

Experimental designs that are commonly used in sensory evaluation are presented. The discussion is structured to avoid a great deal of the confusion that often surrounds this topic. In Section II independent replications of an experiment are distinguished from multiple observations of a single element from a population. Confusing replications with multiple observations, which is a direct result of failing to recognize the population of interest in a study, can lead to the incorrect use of measurement error (obtained from multiple observations) in place of experimental error (obtained through replication) in the statistical analysis of sensory data. As a result, samples are often falsely declared to be significantly different from each other because measurement error is regularly much smaller than experimental error. In Sections III and IV the total experimental design is broken down into two mutually exclusive parts, the blocking structure of the experimental design and the treatment structure of the experimental design, as is done by Milliken and Johnson.[2] The blocking structure of an experimental design is a description of the way the treatments are applied to the experimental material in a study. Blocking structures commonly used in sensory evaluation include the randomized (complete) block design, balanced incomplete-block designs (BIBDs), and split-plot designs. The treatment structure of an experimental design is a description of the association or relationship among the treatments (i.e., products) in a study. Some commonly used treatment structures include the one-way arrangement, factorial experiments, and response surface methodology (or RSM experiments). By maintaining a clear distinction between the blocking structure and the treatment structure of an experimental design it is easy to identify a set of meaningful comparisons among the samples in a study and to determine the appropriate error terms to use in the statistical analysis of any set of data.

A brief discussion of multivariate techniques is presented in Section V. Multivariate approaches can be used to summarize large bodies of data in a manner that uncovers fundamental differences among samples that might otherwise go unnoticed.

II. REPLICATION VS. MULTIPLE OBSERVATIONS

The fundamental intent of a statistical analysis is to generate an accurate and precise estimate of the "experimental error". Experimental error is the unexplainable, natural variability of the population being studied. Experimental error is expressed quantitatively as the variance or as the standard deviation of the population. One measurement taken on one element or unit from a population provides no means for estimating experimental error. In fact, multiple observations of the same element or unit from the population provides no means to estimate experimental error either. The differences among the multiple observations taken on a single unit result from measurement error. Typically, measurement error is much smaller than experimental error. In order to develop a valid estimate of experimental error it is necessary to collect measurements from several units from the population. The measurements taken on different units are called replications. It is the unit-to-unit (or "rep.-to-rep.") differences that contain the information about the variability of the population that is sought (i.e., experimental error, not measurement error).

The objective of most sensory studies is to differentiate products based on differences in the perceived intensities of some attribute or attributes. When only a single sample of each product is evaluated there is no way to estimate experimental error. Often times in such situations measurement error is substituted for experimental error in the statistical analysis of the study. This is a very dangerous mistake because replacing experimental error with the smaller measurement error will often lead to falsely concluding that significant differences exist among the products when, in fact, no such differences exist.

Consider the following artificial example that makes this point quite clearly. Suppose that the objective of the study is to compare the heights of men from three cities in the U.S. In the first case the investigator collects a random sample of 20 men from each city and measures their heights. The data set then consists of 60 height measurements divided into 3 groups by city. In the second case an economy-minded investigator randomly selects 1 man from each city and measures the heights of each of the 3 men repeatedly, 20 times. Again, the data set consists of 60 height measurements divided into 3 groups by city. There is, however, an obvious difference between the two sets of data. In the first case the data set consists of twenty independently replicated height measurements from each city; because each measurement is taken on a different individual from each city. In the second case the data set consists of 20 multiple observations of each of the 3 men's heights. The first investigator is able to develop a valid estimate of the variability of men's heights and can therefore perform valid statistical tests to determine if significant city-to-city differences exist. The second investigator has no way to perform valid statistical tests for city-to-city differences because he has no measure of the man-to-man variability within each city. If this economy-minded investigator uses the measurement error in place of the true experimental error, he will very likely conclude that significant differences exist when in fact they may not. This happens because the multiple observations of height taken on the same person are far less variable than the person-to-person differences. Stated in more technical terms, the first investigator correctly samples the population of interest (i.e., 20 men from each city) and therefore develops a valid estimate of experimental error (i.e., the variability among men's heights). The second investigator samples a totally different population, that is, the population of height measurements taken on a single individual and incorrectly uses the measurement-to-measurement variability as a substitute for person-to-person variability.

If in a taste testing study the contents of one jar of mayonnaise is divided into 20 servings and presented to panelists, or a single preparation of a sweetener solution is poured into twenty cups and served, then the results of the test are equivalent to the second example given above. The variability estimate obtained from the study estimates "measurement" error. It is not a measure of the "product variability" (the valid experimental error) because the "treatments" were not independently replicated. To avoid confusing independent replications of a treatment with multiple observations the sensory analyst must know the legitimate sources of variability of the product's population. If the objective of a study is to compare several brands of a product type that is known to vary day-to-day during production, then single servings of each day's production need to be presented in the sensory study. If an ingredient is known to be extremely uniform, then at the very least, separate preparations of samples with that ingredient should be served to each judge. (For extremely uniform products the major source of variability may well be the preparation-to-preparation differences).

Suggesting that only one sample be taken from each jar of product or that each serving be prepared separately is undeniably more inconvenient than taking multiple observations on a single sample. However, the sensory analyst must compare this inconvenience to the price paid when, for instance, a new product fails in the market because a prototype formulation was falsely declared to be significantly superior to a current formulation based on the evaluation of a single sample of each product.

III. THE BLOCKING STRUCTURE OF AN EXPERIMENTAL DESIGN

The blocking structure of an experimental design is a description of how the treatments are applied to the experimental material. To understand blocking structure two concepts must be understood, the "block" and the "experimental unit". A block is simply a group of homogeneous experimental material.* Theoretically, except for experimental error, any unit within a block will yield the same response to the application of a given treatment. The level of the response to a given treatment may vary from block to block but it is always assumed that the difference between any two treatments applied to units from the same block is the same for all blocks. The experimental material within a block is divided into small groups called "experimental units". An experimental unit is that portion of the total experimental material to which a treatment can be (or is) independently applied. The key word in this definition is "independently".

In a taste test the experimental material is a large group of tastings. The tastings are often arranged into blocks according to judges. Blocking on judges is done in recognition of the fact that, due to differing thresholds for instance, judges may use different parts of the rating scale to express their taste perceptions. It is assumed, however, that the size of the perceived difference between any two samples is the same from judge to judge. Within each judge (i.e., block) the separate tastings are the experimental units. The treatments, which can be thought of as products at this point, must be independently applied at each tasting. This is done to avoid confusing replications with multiple observations of the same sample. Only through the independent applications (i.e., replications) of the treatments can legitimate estimates of experimental error be developed, and thus, legitimate tests of hypotheses be performed.

The simplest blocking structure is the completely randomized design (CRD). In a CRD all of the experimental material is homogeneous, that is, a CRD consists of one large block of experimental units. CRDs are seldom used in sensory evaluation because judge-to-judge differences are known to exist and in most sensory studies the number of products is too great for a single judge to perform replicated evaluations of all of the products without experiencing some change in acuity (thus violating the assumption of block homogeneity). More elaborate blocking structures are required in sensory studies. The remainder of this section is devoted to the discussion of three of the most commonly used blocking structures in sensory evaluation.

A. Randomized (Complete) Block Designs

If the number of samples is sufficiently small so that sensory fatigue is not a concern then a randomized (complete) block design is appropriate. Panelists are the "blocks"; samples are the "treatments". Each panelist evaluates (either by rating or ranking) all of the samples (thus the term "complete block").

A randomized block design is effective when the sensory analyst is confident that the panelists are consistent in rating the samples but recognizes that panelists might use different

---

* Many terms used in the area of experimental design originated from agricultural experiments. "Block" originally meant a block of land in a field that had the same fertility (or that received the same level of irrigation) throughout, while other "blocks" of land in the field might have had different fertility (or irrigation) levels.

Table 1
DATA TABLE FOR A RANDOMIZED (COMPLETE) BLOCK DESIGN

Blocks (judges)	Samples					Row total
	1	2	• • •	t		
1	X_{11}	X_{12}	• • •	X_{1t}		$X_{1\cdot} = \sum_{j=1}^{t} X_{1j}$
2	X_{21}	X_{22}	• • •	X_{2t}		$X_{2\cdot} = \sum_{j=1}^{t} X_{2j}$
•	•	•		•	•	
•	•	•		•	•	
•	•	•		•	•	
b	X_{b1}	X_{b2}	• • •	X_{bt}		$X_{b\cdot} = \sum_{j=1}^{t} X_{bj}$
Column total	$X_{0.1} = \sum_{i=1}^{b} X_{i1}$	$X_{0.2} = \sum_{i=1}^{b} X_{i2}$	• • •	$X_{.t} = \sum_{i=1}^{b} X_{it}$		$X_{..} = \sum_{i=1}^{b} \sum_{j=1}^{t} X_{ij}$

Table 2
ANOVA TABLE FOR RANDOMIZED BLOCK DESIGNS USING RATINGS

Source of variability	Degrees of freedom	Sum of squares	Mean square	F
Total	$bt - 1$	SS_T		
Blocks (judges)	$b - 1$	SS_J		
Samples	$t - 1$	SS_S	$MS_S = SS_S/(t - 1)$	MS_S/MS_E
Error	$df_E = (b - 1)(t - 1)$	SS_E	$MS_E = SS_E/df_E$	

parts of the scale to express their perceptions. The analysis applied to data from a randomized block design takes into account this type of judge-to-judge difference, yielding a more accurate estimate of experimental error and thus more sensitive tests of hypotheses than would otherwise be available.

Independently replicated samples of the test products are presented to the panelists in a randomized order (using a separate randomization for each panelist). The data obtained from the panelists' evaluation of the samples can be arranged in a two-way table as in Table 1.

1. Randomized Block Analysis of Ratings

Data in the form of ratings from a randomized block design are analyzed by ANOVA. The form of the ANOVA table appropriate for a randomized block design is presented in Table 2. The null hypothesis being tested is that the mean ratings for all of the samples are equal (H_0: $\mu_i = \mu_j$ for all $i \neq j$) vs. the alternative hypothesis that the mean ratings of at least two of the samples are different (H_a: $\mu_i \neq \mu_j$ for some pair of samples i and j, $i \neq j$). If the value of the F-statistic calculated as in Table 2 exceeds the critical value of an F with $(t - 1)$ and $(b - 1)$ $(t - 1)$ degrees of freedom (Table T6), then the null hypothesis is rejected in favor of alternative hypothesis. Multiple comparison procedures are then applied to determine which of the samples have significantly different average ratings (see Section III.D, Equation 5 or 9).

2. Randomized Block Analysis of Rank Data

If the data from a randomized block design are in the form of ranks then a nonparametric analysis is performed using a Friedman-type statistic. The data are arranged as in Table 1, but instead of ratings, each row of the table contains the ranks assigned to the samples by each judge. The row totals, at the right of Table 1 are each equal to $t(t + 1)/2$. The column totals at the bottom of Table 1 are the rank sums associated with the samples.

The Friedman-type statistic for rank data, that takes the place of the F-statistic in the analysis of ratings, is

$$T = ([12/bt(t + 1)] \sum_{j=1}^{t} X_{\cdot j}^2) - 3b(t + 1) \tag{1}$$

where b = the number of panelists, t = the number of samples, and $X_{\cdot j}$ = the rank sum corresponding to sample j (i.e., the column total for sample j in Table 1). The "dot" in $X_{\cdot j}$ indicates that summing has been done over the index replaced by the "dot", that is,

$$X_{\cdot j} = \sum_{i=1}^{b} X_{ij}.$$

The test procedure is to reject the null hypothesis (H_0: $\mu_i = \mu_j$ for all $i \neq j$) at the α level of significance if the value of T in Equation 1 exceeds $\chi^2_{\alpha, (t-1)}$; and to accept H_0 otherwise, where $\chi^2_{\alpha, (t-1)}$ is the upper-α percentile of the χ^2 distribution with $(t - 1)$ degrees of freedom (Table T3). The procedure assumes that a relatively large number of panelists participate in the study. It is reasonably accurate for studies involving 12 or more panelists. Multiple comparison procedures for rank data, used to determine which samples differ significantly, are described in Section III.D, Equation 7 or 11.

If the panelists are permitted to assign equal ranks or ties to the samples then a slightly more complicated form of the test statistic T' must be used (see Hollander and Wolfe[3]). Assign the average of the tied ranks to each of the samples that could not be differentiated. For instance, in a four-sample test if the middle two samples (normally of ranks 2 and 3) could not be differentiated, then assign both of the samples the average rank of 2.5 (i.e., $(2 + 3)/2 = 2.5$). Replace T in Equation 1 with

$$T' = \frac{12 \sum_{j=1}^{t} (K_{\cdot j} - G/t)^2}{bt(t + 1) - [1/(t - 1)] \sum_{i=1}^{b} [(\sum_{j=1}^{g_i} t_{i,j}^3) - t]} \tag{2}$$

where $G = bt(t + 1)/2$, g_i is the number of tied groups in block i and $t_{i,j}$ is the size of the jth tied group in block i (i.e., the number of samples in the tied group). Nontied samples are each counted as a separate group of size $t_{i,j} = 1$. Only the blocks in which ties occur need to be considered in the calculation of the second term in the denominator of Equation 2.

B. Balanced Incomplete-Block Designs

When, due to sensory fatigue, it is unreasonable to expect the panelists to evaluate and provide reliable data on all the samples in a study an incomplete-block design should be used. With incomplete-block designs the panelists evaluate only a portion of the total number of samples. Incomplete-block designs provide the means for sensory analysts to obtain consistent, reliable data from their panelists when the total number of samples in the study exceeds the number that can be evaluated before sensory fatigue sets in.

Table 3
DATA TABLE FOR A BALANCED INCOMPLETE-BLOCK DESIGN
($t = 7, k = 3, b = 7, r = 3, \lambda = 1, p = 1$)

Block	Sample 1	2	3	4	5	6	7	Block total
1	X	X		X				B_1
2		X	X		X			B_2
3			X	X		X		B_3
4				X	X		X	B_4
5	X				X	X		B_5
6		X				X	X	B_6
7	X		X				X	B_7
Treatment total	R_1	R_2	R_3	R_4	R_5	R_6	R_7	G

Note: B_i = (sum of entries in row i), R_j = (sum of entries in column j), G = (grand total of the data).

There are several classes of incomplete-block designs. The most widely used class is the balanced incomplete-block designs (BIBs or BIBDs). In a BIB design each block contains the same number of treatments (i.e., samples). Notationally, the total number of samples is denoted by t and the number of samples appearing in each block by k, where k<t and both are integers. In addition, every pair of samples occurs an equal number of times in a BIB design (denoted notationally by λ). These two requirements define "balanced" in BIB designs.

The number of blocks required to complete a single repetition of a BIB design is denoted by b where b $\geq$ t and an integer. Each sample appears r times in a single repetition of the design (i.e., every b blocks). Table 3 illustrates a typical BIB layout. A list of BIB designs, such as the one presented by Cochran and Cox[4] is very helpful in selecting a specific design for a study.

In order to obtain a sufficiently large number of total replications per sample the entire BIB design (b blocks) may have to be repeated several times. The number of repeats or repetitions of the fundamental design is denoted by p. The total number of blocks is then pb, yielding a total of pr replications for every sample and a total of pλ for the number of times every pair of samples occur in the total BIB design.

Because of their balanced nature BIB designs provide equally precise estimates of all treatment effects, that is, the standard error of each sample mean is the same. Also, pair-wise comparisons of any two treatments are equally precise. Experience with 9-point category scales and unstructured line scales has shown that the total number of replications (pr) should be at least 18 in order to yield sufficiently precise estimates of the sample means. This is a rule-of-thumb, suggested only to provide a starting point for determining how many panelists are required for a study. The total number of replications needed to ensure that meaningfully large differences among the samples are declared statistically significant is influenced by many factors: the products, panelist acuity, level of training, etc. Only trial and error can answer the question of how many replications are needed for any given study.

There are two general approaches for administering a BIB design in a sensory study. (1) If the number of blocks is relatively small (four or five, say), it may be possible to have a small number of panelists (p in all) return several times until each panelist has completed an entire repetition of the design. The order of presentation of the blocks should be ran-

Table 4
ANOVA TABLES FOR BALANCED INCOMPLETE-BLOCK DESIGNS

Source of variability	Degrees of freedom	Sum of square	Mean square	F
A. Each of p Panelists Evaluates All b Blocks				
Total	tpr − 1	SS_T		
Panelists	p − 1	SS_P		
Blocks (within panelists)	p(b − 1)	$SS_{B(P)}$		
Samples (adj. for blocks)	t − 1	SS_S	$MS_S = SS_S/(t − 1)$	MS_S/MS_E
Error	df_E = tpr − t − pb + 1	SS_E	$MS_E = SS_E/df_E$	
B. Each of pb Panelists Evaluates One Block				
Total	tpr − 1	SS_T		
Blocks	pb − 1	SS_B		
Samples (adj. for blocks)	t − 1	SS_S	$MS_S = SS_S/(t − 1)$	MS_S/MS_E
Error	df_E = tpr − t − pb + 1	SS_E	$MS_E = SS_E/df_E$	

domized for each panelist separately. (2) For large values of b, the normal practice is to call upon a large number of panelists (pb in all) and to have each evaluate the samples in a single block. The block of samples that a particular panelist receives should be assigned at random. The order of presentation of the samples within each block should be randomized in all cases. Also, the samples of the test products should be independently replicated at each serving.

1. BIB Analysis of Ratings

ANOVA is used to analyze BIB data in the form of ratings (see Table 4). As in the case of a randomized (complete) block design, the total variability is partitioned into the separate effects of block, treatments (samples), and error. However, the formulas used to calculate the sum of squares in a BIB analysis are more complicated than for a randomized (complete) block analysis. The sensory analyst should ensure that the statistical package used to perform the analysis is capable of handling a BIB design. Otherwise a program specifically developed to perform the BIB analysis is required.

The form of the ANOVA used to analyze BIB data depends on how the design is administered. If each panelist evaluates every block in the fundamental design, then the "panelist effect" can be partitioned out of the total variability (see Table 4A). If each panelist evaluates only one block of samples, then the panelist effect is confounded (or mixed-up) with the block effect. It is not pooled into the error term (see Table 4B).

If the F-statistic in Table 4 exceeds the critical value of an F with the corresponding degrees of freedom then the null hypothesis assumption of equivalent mean ratings among the samples is rejected. Multiple comparison procedures for BIB designs are discussed in Section III.D, Equation 6 or 10.

2. BIB Analysis of Rank Data

A Friedman-type statistic is applied to rank data arising from a BIB design. The form of the test statistic is

$$T = (12/p\lambda t(k + 1)) \sum_{j=1}^{t} R_j^2 − 3(k + 1)pr^2/\lambda \tag{3}$$

where t, k, r, λ, and p are defined above and R_j is the rank sum associated with the jth sample (i.e., the rank sum of sample j in the last row of Table 3) (see Durbin[5]). Tables of critical values of T in Equation 3 are available for selected combinations of t = 3 to 6, k = 2 to 5, and p = 1 to 7 (see Skillings and Mack[6]). However, in most sensory testing situations the total number of blocks exceeds the values in the tables. For these situations, the test procedure is to reject the assumption of equivalency among the samples if T in Equation 3 exceeds the upper-α critical value of the χ^2 distribution with (t − 1) degrees of freedom (i.e., reject H_0 if the $T \geq \chi^2_{\alpha, (t-1)}$, Table T3). Multiple comparison procedures are presented in Section III.D, Equation 8 or 12.

C. Split-Plot Designs

In randomized block and BIBDs, panelists are treated as a blocking factor, that is, it is assumed that the panelists are an identifiable source of variability that should be compensated for in the design of a sensory study. In ANOVA the effects of blocking factors (e.g., judges) and treatment factors (e.g., products) are assumed to be additive. In practice this assumption implies that while panelists may use different parts of the sensory rating scales to express their perceptions, the size and direction of the differences among the samples are perceived and reported in the same way by all of the panelists. Of course, the data actually collected in a study diverge slightly from the assumed pattern due to experimental error. Another way of stating this assumption is that there is no "interaction" between blocks and treatments (e.g., judges and samples) in a randomized block or BIB design. This is one of the fundamental assumptions of ANOVA. For a group of highly trained, motivated, and "calibrated" panelists the assumption of no interaction between block and treatments is reasonable. However, during training for instance, the sensory analyst may doubt the validity of this assumption. Split-plot designs are used to determine if a judge-by-sample interaction is present.

In split-plot designs judges are treated at a second treatment factor along with the samples. A group of b panelists are presented with t samples (in a separately randomized order for each panelist) in each of at least two panels (p ≥ 2). The p panels are the blocks or "replicates" of the experimental design. First, the panelists are randomly assigned a place in the panel (normally by their arrival times). Then the t samples are presented to each of the b panelists at each of the p panels. Due to the sequential nature of the randomization scheme, where first one treatment factor (judges) is randomized within replicates and then a second treatment factor (samples) is randomized within the first treatment factor (i.e., judges), a split-plot design is appropriate.

1. Split-Plot Analysis of Ratings

A special form of ANOVA is used to analyze data from a split-plot design (see Table 5). The panelist effect is called the "whole-plot effect". Samples and the panelist-by-sample interaction are called the "subplot effects". Separate error terms are used to test for the significance of whole-plot and subplot effects (because of the sequential nature of the randomization scheme discussed above).

The whole-plot error term (Error(A) in Table 5) is calculated in the same way as a panel-by-panelist interaction term would be, if one existed. The F_1 statistic in Table 5 is used to test for a significant panelist effect. If the value of F_1 is larger than the upper-α critical value of the F distribution with (b − 1) and (p − 1)(b − 1) degrees of freedom, then it is concluded that there are significant differences in the average values of the panelists' responses. A significant panelist effect confirms that the panelists are using different parts of the rating scale to express their perceptions.

The F_2 and F_3 statistics in Table 5 are used to test for the significance of the subplot effects, samples, and the judge-by-sample interaction, respectively. The denominator of both

Table 5

ANOVA TABLE FOR SPLIT-PLOT DESIGNS USING RATINGS

Source of variability	Degrees of freedom	Sum of squares	Mean square	F
Total	$pb(t - 1)$	SS_T		
Panel	$p - 1$	SS_P		
Judges	$b - 1$	SS_J	$MS_J = SS_J/(b - 1)$	$F_1 = MS_J/MS_{E(A)}$
Error(A)	$df_{E(A)} = (p - 1)(b - 1)$	$SS_{E(A)}$	$MS_{E(A)} = SS_{E(A)}/df_{E(A)}$	
Samples	$t - 1$	SS_S	$MS_S = SS_S/(t - 1)$	$F_2 = MS_S/MS_{E(B)}$
Judge-by-sample	$df_{JS} = (b - 1)(t - 1)$	SS_{JS}	$MS_{JS} = SS_{JS}/df_{JS}$	$F_3 = MS_{JS}/MSE_{(B)}$
Error(B)	$df_{E(B)} = b(p - 1)(t - 1)$	$SS_{E(B)}$	$MS_{E(B)} = SS_{E(B)}/df_{E(B)}$	

F_2 and F_3 is the subplot error term MS_B. If F_3 exceeds the upper-α critical value of the F distribution with $(b - 1)(t - 1)$ and $b(p - 1)(t - 1)$ degrees of freedom, then a significant judge-by-sample interaction exists. The significance of the interaction indicates that the judges are expressing their perceptions of the differences among the samples in different ways. Judge-by-sample interactions result from insufficient training, confusion over the definition of the attribute being evaluated, or lack of familiarity with the rating technique. When a significant judge-by-sample interaction exists it is meaningless to examine the overall sample effect (tested by F_2 in Table 5) because the presence of the interaction indicates that the reported nature or pattern of the differences among the samples depends on which judge or judges are being considered.

If F_3 is not significant but F_2 is, then an overall sample effect is present. Significant differences among the samples can be identified by using the multiple comparison procedures presented in Equations 5 and 9 in the next section.

D. Multiple Comparison Procedures

An F-statistic in ANOVA or a T-statistic in a Friedman analysis is used to determine if there are any significant differences among the samples in a study. Both are overall tests. Neither provides any information about which samples are significantly different from each other. Multiple comparison procedures are used to determine which samples in a study differ significantly.

The method for applying the multiple comparison procedures discussed in this section is straightforward. First, the value of the multiple comparison is calculated. Next, the differences between every pair of sample means or rank sums are calculated. Whenever the magnitude of the difference between two means or rank sums exceeds the value of the multiple comparison the two corresponding samples are declared to be significantly different.

There are two classes of multiple comparison procedures. The first class controls the comparison-wise error rate, that is, the type I error (of size α) applies each time a comparison of means or rank sums is made. Procedures that control the comparison-wise error rate are called one-at-a-time multiple comparison procedures. The second class controls the experiment-wise error rate, that is, the type I error applies to all of the comparisons among means or rank sums simultaneously. Procedures that control the experiment-wise error rate are called simultaneous multiple comparison procedures.

1. One-at-a-Time Multiple Comparison Procedures

Fisher's LSD (least significant difference) is a one-at-a-time multiple comparison procedure. Fisher's LSD should only be applied if the overall test statistic, F or T, is significant. The general form of Fisher's LSD is

$$LSD = t_{\alpha/2,df_E} \sqrt{MS_E} \sqrt{(1/n_1) + (1/n_2)} \tag{4}$$

where $t_{\alpha/2,df_E}$ is the upper-$\alpha/2$ critical value of a student's t distribution with df_E degrees of freedom (i.e., the degrees of freedom for error from the ANOVA), MS_E is the mean square for error from the ANOVA, and n_1 and n_2 are the sample sizes used to calculate the two means being compared. If $n_1 = n_2$ then the LSD simplifies to

$$LSD = t_{\alpha/2,df_E} \sqrt{2MS_E/n} \tag{5}$$

where n is the common sample size. For randomized (complete) block designs the value of n in Equation 5 is b, the number of blocks (typically, the number of judges) in the study. For split-plot designs the value of n in Equation 5 is pb, the number of panels times the number of blocks. (Also, for split-plot designs, MS_E is the mean square for Error(B) from the ANOVA.) The form of Fisher's LSD for BIB designs is more complicated than Equations 4 or 5 due to the fact that not all samples appear in each block of the design. Fisher's LSD for BIB designs has the form

$$LSD = t_{\alpha/2,df_E} \sqrt{2MS_E/pr} \sqrt{[k(t - 1)]/[(k - 1)t]} \tag{6}$$

where t is the number of treatments, k is the number of samples appearing in each block, r is the number of times each sample is evaluated in the fundamental design (i.e., in one repetition of b blocks), and p is the number of times the fundamental design is repeated. MS_E and $t_{\alpha/2,df_E}$ are as defined above.

A nonparametric analog to Fisher's LSD for rank sums from a randomized (complete) block design is

$$LSD_{rank} = z_{\alpha/2}\sqrt{bt(t + 1)/6}$$
$$= t_{\alpha/2,\infty}\sqrt{bt(t + 1)/6} \tag{7}$$

Similarly, a nonparametric analog to Fisher's LSD for rank sums from a BIB design is

$$LSD_{rank} = z_{\alpha/2}\sqrt{p(k + 1)(rk - r + \lambda)/6} \tag{8}$$

2. Simultaneous Multiple Comparison Procedures

Tukey's HSD (honestly significant difference) is a simultaneous multiple comparison procedure. Tukey's HSD can be applied regardless of the outcome of the overall test for differences among the samples. The general form of Tukey's HSD for the equal sample-size case for ratings data is

$$HSD = q_{\alpha,t,df_E}\sqrt{MS_E/n} \tag{9}$$

where q_{α,t,df_E} is the upper-α critical value of the studentized range distribution with df_E degrees of freedom (Table T14) for comparing t sample means. As with the LSD, df_E and MS_E are the degrees of freedom and the mean square for error from the ANOVA, respectively (Error(B) in the split-plot ANOVA); n is the sample size common to all the means being compared. For randomized (complete) block designs n = b; for split-plot designs n = pb. Tukey's HSD for BIB designs has the form

$$HSD = q_{\alpha,t,df_E}\sqrt{MS_E/pr} \sqrt{k(t - 1)/(k - 1)t} \tag{10}$$

The nonparametric analog to Tukey's HSD for rank sums is

$$HSD_{rank} = q_{\alpha,t,\infty}\sqrt{bt(t + 1)/12} \qquad (11)$$

for randomized (complete) block designs, and

$$HSD_{rank} = q_{\alpha,t,\infty}\sqrt{p(k + 1)(rk - r + \lambda)/12} \qquad (12)$$

for BIB designs.

IV. THE TREATMENT STRUCTURE OF AN EXPERIMENTAL DESIGN

In all of the experimental designs discussed thus far the treatments or products have been viewed as a set of qualitatively distinct objects, having no particular association among themselves. Such designs are said to have a one-way treatment structure. Designs with one-way treatment structures commonly occur at the end of a research program when the objective is to decide which product should be selected for further development.

In many experimental situations, however, the focus of the research is not on the specific samples but rather on the effects of some factor or factors that have been applied to the samples. For instance, a researcher may be interested in the effects that different flour and sugar have on the flavor and texture of a specific cake recipe, or he may be interested in the effects that cooking time and temperature have on the flavor and appearance of a prepared meat. In situations such as these there are specific treatment structures available that provide highly precise and comprehensive comparisons of the effects of the factors on the samples, while at the same time minimize the total amount of experimental material required to perform the study.

Two such "multi-way" treatment structures are discussed in this section. They are the factorial treatment structure (often called factorial experiments) and the response surface treatment structure (often called response surface methodology or RSM).

A. Factorial Treatment Structures

Researchers are often interested in studying the effects that two or more factors have on a set of responses. Factorial treatment structures are the most efficient means to perform such studies. In a factorial experiment specific levels for each of several factors are defined. A single replication of a factorial experiment consists of all possible combinations of the levels of the factors. For example, a brewer may be interested in comparing the effects of two kettle boiling times on the hop aroma of his beer. Further, the brewer is currently using two varieties of hops and he is not sure if the two varieties of hops respond similarly to changes in kettle boiling times. Combining the two levels of the first factor, kettle boiling time, with the two levels of the second factor, variety of hops, yields four distinct treatment combinations that form a single replication of a factorial experiment (see Table 6).

The levels of the factors in a factorial experiment may differ quantitatively (e.g., boiling time) or qualitatively (e.g., variety of hops). Any combination of quantitative and qualitative factors may be run in the same factorial experiment.

The effect of a factor is the change (or difference) in the response that results from a change in the level of the factor. The effects of individual factors are called "main effects". For example, if the entries in Table 6 represent the average hop aroma rating of the four beer samples, the main effect due to boiling time is

$$\frac{(T_{1A} - T_{2A}) + (T_{1B} - T_{2B})}{2} \qquad (13)$$

Table 6
FACTORIAL TREATMENT
STRUCTURE FOR TWO
FACTORS EACH HAVING TWO
LEVELS

Kettle boiling time	Hop variety	
	A	B
Low (1)	$T_{1A} = 6$	$T_{1B} = 13$
High (2)	$T_{2A} = 12$	$T_{2B} = 7$

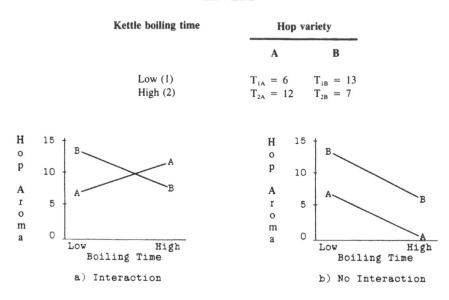

FIGURE 1. Plots of the mean response illustrating an interaction (a) and no interaction (b) between the factors in the study.

Similarly, the main effect due to variety of hops is

$$\frac{(T_{1A} - T_{1B}) + (T_{2A} - T_{2B})}{2} \tag{14}$$

In some studies the effect of one factor depends on the level of a second factor. When this occurs there is said to be an interaction between the two factors. Suppose for the beer brewed with hop variety A that the hop aroma rating increased when the kettle boiling time was changed from its low to its high level, but that hop aroma decreased for the same change in boiling time when the beer is brewed with hop variety B (see Table 6). There is an interaction between kettle boiling time and variety of hops because the effect of boiling time depends on which variety of hops is being used.

Graphs can be used to illustrate interactions. Figure 1a illustrates the interaction between boiling time and variety. The points on the graph are the average hop aroma ratings of the four experimental points presented in Table 6. Note the lack of parallelism between the two lines. This indicates an interaction between the two factors. If there is no interaction between the two factors the lines are nearly parallel (deviating only due to experimental error) as in Figure 1b. Researchers must be very cautious in interpreting main effects in the presence of interactions. Consider the data in Table 6 that illustrate the "boiling-time by hop-variety" interaction. Applying Equation 13 yields an estimated main effect due to boiling time of ((6 − 12) + (13 − 7))/2 = 0, which suggests that there is no effect due to boiling time. However, Figure 1a shows quite clearly that at each level of the hop-variety factor there is a substantial effect due to boiling time. Because the separate variety effects are opposites

Table 7

ANOVA TABLE FOR A FACTORIAL EXPERIMENT

Source of variation	Degrees of freedom	Sum of squares	Mean square	F
Total	$rab - 1$	SS_T		
A	$a - 1$	SS_A	$MS_A = SS_A/(a - 1)$	$F_A = MS_A/MS_E$
B	$b - 1$	SS_B	$MS_B = SS_B/(b - 1)$	$F_B = MS_B/MS_E$
AB	$df_{AB} = (a - 1)(b - 1)$	SS_{AB}	$MS_{AB} = SS_{AB}/df_{AB}$	$F_{AB} = MS_{AB}/MS_E$
Error	$df_E = ab(r - 1)$	SS_E	$MS_E = SS_E/df_E$	

Note: Factor A has "a" levels, Factor B has "b" levels, and the entire factorial experiment is replicated "r" times. The samples are prepared according to a completely randomized blocking structure.

they cancel each other out in calculating the main effect due to boiling time. In the presence of an interaction, the effect of one factor can only be meaningfully studied by holding the level of the second factor fixed.

Researchers sometimes use an alternative to factorial treatment structures, called one-at-a-time treatment structures, in the false belief that they are economizing the study. Suppose in the beer brewing example that the brewer had only prepared three samples: the low boiling time/variety A point T_{1A}, the low boiling time/variety B point T_{1B}, and the high boiling time/variety B point T_{2B}. (The high boiling time/variety A treatment combination T_{2A} is omitted.) Since only three samples are prepared it would appear that the one-at-a-time approach is more economical than the full factorial approach. However, consider the precision of the estimates of the main effects in the presence of experimental error. Only one difference due to boiling time is available to estimate the main effect due to boiling time in the one-at-a-time study (i.e., $T_{1B} - T_{2B}$). The same is true for the variety effect (i.e., $T_{1A} - T_{1B}$). In the factorial treatment structure two differences are available for estimating each effect, as shown in Equations 13 and 14. The entire one-at-a-time experiment would have to be replicated twice, yielding six experimental points, to obtain estimates of the main effects that are as precise as those obtained from the four points in the factorial experiment.

Another advantage that factorial treatment structures have over one-at-a-time experiments is the ability to estimate interactions. If the high temperature/variety A observation $T_{2A} = 12$ is omitted from the data in Table 6 (as in the one-at-a-time study, above), one would observe that beer brewed at the high boiling time has less hop aroma than beer brewed at the low boiling time and that beer brewed with hop variety A has less hop aroma than beer brewed with hop variety B. The most obvious conclusion would be that beer brewed at the high boiling time using hop variety A would have the least hop aroma of all. Both the data in Table 6 and the plot of the interaction in Figure 1a show that this could be a terribly erroneous conclusion.

To summarize, factorial treatment structures are economical, provide a means for estimating interactions, and permit researchers to study the effects of several factors over a broad range of experimental conditions.

The recommended procedure for applying factorial treatment structures in sensory evaluation is as follows. Prepare at least two independent replications of the full factorial experiment. Submit the resulting samples for panel evaluation using the appropriate blocking structure as described in Section III. Take the mean responses from the analysis of the panel data and use them as raw data in an ANOVA. The output of the ANOVA includes tests for main effects and interactions among the experimental factors (See Table 7). This procedure avoids confusing the measurement error, obtained from the analysis of the panel data, with the true experimental error, which can only be obtained from the differences among the independently replicated treatment combinations.

B. Response Surface Methodology

The treatment structure known as response surface methodology (RSM) is essentially a designed regression analysis (see Montgomery[7] and Giovanni[8]). Unlike factorial treatment structures, where the objective is to determine if (and how) the factors influence the response, the objective of an RSM experiment is to develop a regression equation that predicts the value of a response variable (called the dependent variable) based on the controlled values of the experimental factors (called the independent variables). All of the factors in an RSM experiment must be quantitative.

RSM treatment structures provide an economical way to predict the value of one or several responses over a range of values of the independent variables. A set of examples (i.e., experimental points) is prepared under the conditions specified by the selected RSM treatment structure. The samples are analyzed by a sensory panel and the resulting average responses are submitted to a stepwise regression analysis. The regression procedure yields a predictive equation that relates the value of the response(s) to the values of the independent variables. The predictive equation can be depicted graphically in what is called a response surface "contour plot", such as the one shown in Figure 2. Contour plots are easy to interpret. They allow the researcher to determine the predicted value of the response at any point inside the experimental region without requiring that a sample be prepared at that point.

Several classes of treatment structures can be used as RSM experiments. The most widely used class, discussed here, is very similar to the treatment structure of a factorial experiment. The main portion of the treatment structure consists of all possible combinations of the low and high levels of the independent variables. (In a two-factor RSM experiment this portion consists of the four points: (low,low), (low,high), (high,low), and (high,high).) The main portion of the treatment structure is augmented by a center point (i.e., the point where all of the factors take on their average values, (low + high)/2). Typically the center point is replicated several times (not less than three) to provide an independent estimate of experimental error (See Figure 3). The regular practice in an RSM experiment is to assign the low levels of all the factors the coded value of -1; the high levels are all assigned the coded value of $+1$; and the center point is assigned the coded value of 0.

The treatment structure of an RSM experiment depicted in Figure 3 is called a first-order RSM experiment. The full regression equation that can be fit by the treatment structure has the form,

$$Y = B_0 + B_1X_1 + B_2X_2 + \ldots + B_kX_k$$
$$+ B_{12}X_1X_2 + B_{13}X_1X_3 + \ldots + B_{k-1,k}X_{k-1}X_k \tag{15}$$

where B_i is the coefficient of the regression equation to be estimated and the X_i the coded level of the k factors in the experiment. First-order RSM experiments are used to depict the general trends of the effects of the independent variables (X_i) on the dependent variable (Y). First-order models are used early in a research program to identify the direction in which to shift the levels of the independent variables to affect a desirable change in the dependent variable (e.g., increase desirable response or decrease an undesirable response).

First-order models may not be able to adequately predict the response if there is a complex relationship between the dependent variable and the independent variables. A second-order RSM treatment structure is required for these situations. The full regression model that can be fit to a second-order RSM treatment structure has the form:

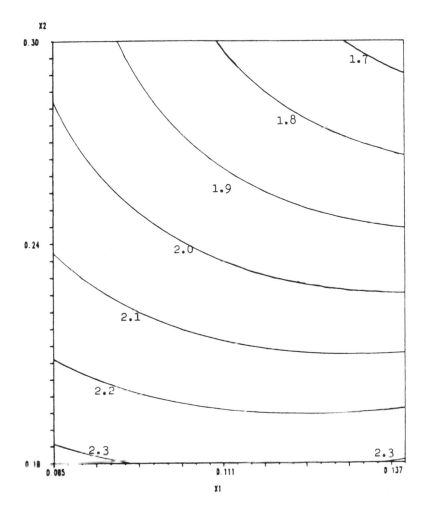

FIGURE 2. Contour plot for a two-factor response surface experiment. $R_1 = 3.45 - 1.77X_1 - 5.83X_2 + 41.77X_1^2 + 13.89_2^2 - 44.25X_1X_2$.

$$Y = B_0 + B_1X_1 + B_2X_2 + \ldots + B_kX_k$$
$$+ B_{11}X_1^2 + B_{22}X_2^2 + \ldots + B_{kk}X_k^2$$
$$+ B_{12}X_1X_2 + B_{13}X_1X_3 + \ldots + B_{k-1,k}X_{k-1}X_k \quad (16)$$

The addition of the squared terms in the model allows the predicted response surface to "bend" and "flex" more than a first-order model, thus resulting in an improved prediction of complex relationships.

An often used class of second-order RSM experiments is the central-composite, rotatable treatment structures. Central-composite, rotatable treatment structures are developed by adding a set of axial or "star" points to a first-order RSM treatment structure. There are 2k axial points in a k-factor RSM experiment. Using the normal -1, 0, $+1$ coding for the factor levels, the axial points are $(\pm\alpha,0,\ldots,0)$, $(0,\pm\alpha,0,\ldots,0)$, $\ldots$, $(0,0,\ldots,0,\pm\alpha)$, where α is the distance the axial point is from the center of the experimental region (i.e., the center point). The value of α is $(F)^{1/4}$, where F is the number of non-center (or factorial) points in the first-order treatment structure. For example, in a two-

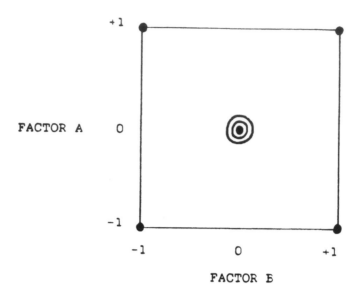

FIGURE 3. A two-factor first-order RSM treatment structure. The figure illustrates the arrangement of factorial and center points in an RSM treatment structure with two independent variables (i.e., factors) that permit estimation of a first-order regression model as in Equation 15.

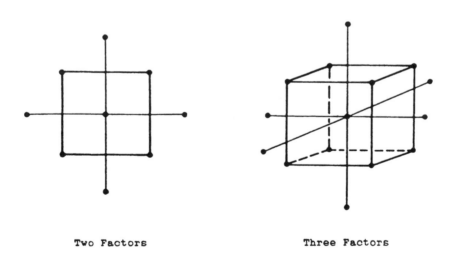

FIGURE 4. Central composite, rotatable RSM treatment structures. The figures illustrate the arrangement of factorial, axial, and center points in RSM treatment structures with two and three independent variables (i.e., factors) that permit estimation of a second-order regression model as in Equation 16.

factor RSM experiment $F = 4$ and $\alpha = (4)^{1/4} = 1.414$. Central-composite, rotatable RSM treatment structures for two and three factor experiments are presented in Figure 4.

Second-order RSM models have several advantages over first-order models. As mentioned above, the second-order models are better able to fit complex relationships between the dependent variable and the independent variables. In addition, second-order models can be used to locate the predicted maximum or minimum value of a response in terms of the levels of the independent variables.

The recommended procedure for performing an RSM experiment is as follows. First, perform a regular BIB analysis of the samples from the RSM treatment structure, ignoring the association among the samples. (The BIB blocking structure is suggested because there are normally too many samples in an RSM treatment structure to evaluate all of them at one sitting. If, however, it is possible to evaluate all of the samples in one block, then a randomized (complete) block design can be used.) The only output of interest from the BIB analysis is the set of adjusted sample means. The significance (or lack of significance) of the overall test statistic is of no interest. The second step of the procedure is to submit the sample means to a stepwise regression analysis in order to develop the predictive equation that relates the value of the response to the level of the experimental factors. The predictive equation is then used to generate a contour plot that provides a graphical depiction of the effects of the factors on the response. If there is only one response, the region where the response takes on acceptable values (or attains a minimum or maximum value) is apparent in the contour plot. When several responses are being considered the individual contour plots can be overlayed. Hopefully, a region where all of the responses take on acceptable values can be identified.

V. MULTIVARIATE STATISTICAL METHODS

The field of sensory evaluation is ripe for the application of multivariate statistical methods. Sensory analysts regularly collect many responses on each sample they evaluate. It is seldom true that the response obtained for one attribute is independent of the responses obtained for all of the remaining attributes measured. Whenever there is an interrelation or ''correlation'' among responses multivariate analyses are more appropriate than their univariate counterparts (which intrinsically assume that the different responses are independent). There is, however, a major obstacle to applying multivariate techniques to the results of a multi-attribute sensory evaluation. The sample sizes required to perform reliable and sensitive multivariate analyses can be prohibitive. One rule of thumb for determining sample size requirements for multivariate analyses states that the sample size should be at least the square of the number of responses ($n \geq p^2$, where p is the number of responses). If $n = 10$ responses are measured the rule of thumb suggests that the sample size $n \geq 100$! (Note here that n is the number of independent samples, not the number of times a given sample is evaluated in a study.) The restrictive sample size requirements not withstanding, sensory analysts should still be aware of what multivariate methods are available and what each method has to offer in terms of summarizing large sets of sensory data.

Principal components and factor analysis are multivariate techniques that can be used to summarize a set of data containing many responses by a small number of ''derived responses''. The derived responses take into account the correlated nature of the original sensory measurements; 2 or 3 derived responses can often summarize as many as 30 original sensory responses with only a minimum loss of information. The derived responses readily lend themselves to graphical presentation. Often times patterns of similarity or differences among the samples, that would otherwise go unnoticed, reveal themselves clearly in the derived responses.

Discriminant analysis is a multivariate technique that differentiates groups of products based on a function of the many original measurements. Discriminant analysis has been applied quite successfully to the problem of differentiating food products based on their volatile flavor profiles obtained from a gas chromatograph, by Powers and Keith[9] and Powers.[10]

Multivariate analysis of variance (MANOVA) extends the ability to compare the average values of a group of products from one response (as in ANOVA) to many responses simultaneously. Due to the interrelated nature of sensory responses, it is possible that MA-

NOVA would declare samples to be significantly different while univariate ANOVAs performed on each response separately would find no significant differences among the samples.

One further word of warning regarding multivariate methods is necessary. In general, the results of a multivariate analysis apply only to the products evaluated. The results do not necessarily apply to an entire class of products, nor do they necessarily apply for all time. This is particularly true when specific samples are intentionally included in the study (as is often the case) because one of the assumptions underlying a multivariate analysis is that the products evaluated are from a random sample drawn from the population of all products in the class of interest. Intentionally including specific products violates this assumption. This warning is not a general indictment against multivariate methods. Multivariate analyses may well yield the most efficient and meaningful summaries of large bodies of data available. However, it is inappropriate and potentially misleading to expect more from the method than what it is intended to deliver.

Readers interested in further discussions of multivariate statistical methods are encouraged to seek out references on the topic. A good starting point is the discussion of multivariate methods presented by Powers and Moskowitz[11] and Powers.[12]

REFERENCES

1. **Kirk, R. E.**, *Experimental Design: Procedures for the Behavioral Sciences*, Brooks/Cole Publ., Belmont, Calif., 1968.
2. **Milliken, G. A. and Johnson, D. E.**, *Analysis of Messy Data*, Vol. I, Lifetime Learning Publ., Belmont, Calif., 1984.
3. **Hollander, M. and Wolfe, D. A.**, *Nonparametric Statistical Methods*, John Wiley & Sons, New York, 1973, 140.
4. **Cochran, W. G. and Cox, G. M.**, *Experimental Designs*, John Wiley & Sons, New York, 1957.
5. **Durbin, J.**, Incomplete blocks in ranking experiments, *Br. J. Stat. Psychol.*, 4, 85, 1951.
6. **Skillings, H. H. and Mack, G. A.**, On the use of a Friedman type statistic in balanced and unbalanced block designs, *Technometrics*, 23, 171, 1981.
7. **Montgomery, D. C.**, *Design and Analysis of Experiments*, John Wiley & Sons, New York, 1976.
8. **Giovanni, M.**, Response surface methodology and product optimization, *Food Technol.*, 37(11), 41, 1983.
9. **Powers, J. J. and Keith, E. S.**, Stepwise discriminant analysis of gas chromatographic data as an aid of classifying the flavor quality of foods, *J. Food Sci.*, 33, 207, 1968.
10. **Powers, J. J.**, Experiences with subjective/objective correlation, Correlating Sensory Objective Measurements: New Methods for Answering Old Problems, ASTM STP 594, American Society for Testing and Materials, Philadelphia, 1976, 111.
11. **Powers, J. J. and Moskowitz, H. R., Eds.**, Correlating Sensory Objective Measurements: New Methods for Answering Old Problems, ASTM STP 594, American Society for Testing and Materials, Philadelphia, 1976.
12. **Powers, J. J.**, Using general statistical programs to evaluate sensory data, *Food Technol.*, 38(6), 74, 1984.

Chapter 13

GUIDELINES FOR CHOICE OF TECHNIQUE

I. INTRODUCTION

The five tables below are meant as memory joggers. They are not a substitute for study of the individual methods described in this book, but once the methods have become familiar, preferably via practical hands-on testing of most of them, the tables can be used to check whether there might be a better way to attack a given problem. Most of us tend to give preference to a few trusted favorite tests, and perhaps to bend the test objective a bit to allow their use, a dangerous habit.

To avoid it, or get out of it, the authors suggest the following practical steps.

A. Define the Project Objective

Read the text in Chapter 1, then refer to Table 1 to classify the type of project. Review the 13 entries. Write down the project objective, then look up the tests to which the table refers.

B. Define the Test Objective

Four tables are available for this purpose:

- Table 2: Difference tests: Does a sensory difference exist between samples?
- Table 3: Attribute difference tests: How does attribute X differ between samples?
- Table 4: Affective tests: Which sample is preferred? How acceptable is sample X?
- Table 5: Descriptive tests: Rate each of the attributes listed in the scoresheet.

Write down the test objective and list the tests required. Then meet with the project leader and others involved in the project and discuss and refine the design of the tests.

C. Reissue Project Objective and Test Objectives. Revise Test Design

In sensory testing, a given problem frequently requires appreciable thought before the appropriate practical tests can be selected. This is because the *initial conception of the problem* may require clarification. It is not unusual for problem and test objectives to be defined and redefined several times before an acceptable design emerges. Sensory tests are expensive, and they often give results which cannot be interpreted. If this happens, the design may be at fault. Pilot tests are often useful as a means of refining a design. It would, for example, be meaningless to carry out a consumer preference test with hundreds of participants without first having shown that a perceptible difference exists; the latter can be established with 10 or 20 tasters, using a difference test. In another example, islands of opposing preference may exist, invalidating a normal preference test; here, the solution may be a pilot study in which various types of customers receive single-sample acceptability tests.

Table 1
TYPES OF PROBLEMS ENCOUNTERED IN SENSORY ANALYSIS[1]

Type of problem	Tests applicable
1. New Product Development. The product development team needs information on the sensory characteristics and also on consumer acceptability of experimental products as compared with existing products in the market.	All tests in this book
2. Product Matching. Here the accent is on proving that no difference exists between an existing and a developmental product.	Similarity tests, Chapter 6
3. Product Improvement. Step 1: define exactly what sensory characteristics need improvement. Step 2: determine that the experimental product is indeed different. Step 3: confirm that experimental product is liked better than the control.	All difference tests, Table 2; then Affective tests, Table 4
4. Process Change. Step 1: confirm that no difference exists. Step 2: if a difference does exist, determine how consumers view the difference.	Similarity tests, Chapter 6; Affective tests, Table 4
5. Cost Reduction and/or Selection of New Source of Supply. Step 1: confirm that no difference exists. Step 2: if a difference does exist, determine how consumers view the difference.	Similarity tests, Chapter 6; Affective tests, Table 4

Note to 3., 4., and 5. above: if new product is different, Descriptive tests (Table 5) may be useful in order to characterize the difference. If the difference is found to be in a single attribute, Attribute difference tests (Table 3) are the tools to use in further work.

6. Quality Control. Products sampled during production, distribution, and marketing are tested to ensure that they are as good as the standard. Descriptive tests (well-trained panel) can monitor many attributes simultaneously.	Difference tests, Table 2; Descriptive tests, Table 5
7. Storage Stability. Testing of current and experimental products after standard aging tests. Step 1: ascertain when difference becomes noticeable; Step 2: Descriptive tests (well-trained panel) can monitor many attributes simultaneously; Step 3: Affective tests can determine the relative acceptance of stored products.	Difference tests, Table 2; Descriptive tests Table 5; Affective tests, Table 4
8. Product Grading or Rating. Used where methods of grading exist which have been accepted by agreement between producer and user, often with government supervision.	Grading, Chapter 5
9. Consumer Acceptance and/or Opinions. After laboratory screening, it may be desirable to submit product to a central-location or home-placement test to determine consumer reaction. Acceptance tests will indicate whether the current product can be marketed, or improvement is needed.	Affective tests, Table 4; Chapter 9
10. Consumer Preference. Full scale consumer preference tests are the last step before test marketing. Employee preference studies cannot replace consumer tests but can reduce their number and cost whenever the desirability of key attributes of the product is known from previous consumer tests.	Affective tests, Table 4; Chapter 9
11. Panelist Selection and Training. An essential activity for any panel. May consist of (1) interview, (2) sensitivity tests, (3) difference tests, and (4) descriptive tests.	Chapter 10
12. Correlation of Sensory with Chemical and Physical Tests. Correlation studies are needed (1) to lessen the load of samples on the panel by replacing a part of the tests with laboratory analyses; (2) to develop background knowledge of the chemical and physical causes of each sensory attribute.	Descriptive tests, Table 5; Attribute difference tests, Table 3
13. Threshold of Added Substances. Required (1) in trouble shooting (Chapter 7) to confirm suspected source(s) of off-flavor(s); (2) to develop background knowledge of the chemical cause(s) of sensory attributes and consumer preferences.	

Table 2

AREA OF APPLICATION OF DIFFERENCE TESTS; DOES A SENSORY DIFFERENCE EXIST BETWEEN SAMPLES?

The tests in this table are suitable for applications such as:

(1) To determine whether product differences result from a change in ingredients, processing, packaging or storage
(2) To determine whether an overall difference exists, where no specific attribute(s) can be identified as having been affected
(3) To select and train panelists and to monitor their ability to discriminate between test samples

Test	Areas of Application
1. Triangle test	Two samples not visibly different. One of the most-used difference tests. Statistically efficient but somewhat affected by sensory fatigue and memory effects. Generally 20—40 subjects, can be used with as few as 5—8 subjects. Brief training required.
2. Two-Out-of-Five test	Two samples without obvious visible differences. Statistically highly efficient but strongly affected by sensory fatigue, hence use limited to visual, auditory, and tactile applications. Generally 8—12 subjects, can be used with as few as 5. Brief training required.
3. Duo-Trio tests	Two samples not visibly different. Test has low statistical efficiency but is less affected by fatigue than triangular test. Useful where product well known to subjects can be employed as the reference. Generally 30 or more subjects, can be used with as few as 12—15. Brief training required.
4. Simple Difference test	Two samples not visibly different. Test has low statistical efficiency but is suitable for samples of strong or lingering flavor, samples which need to be applied to the skin in half face tests, and samples which are very complex stimuli and therefore confusing to the subjects. Generally 30 or more subjects, can be used with as few as 12—15. Brief training required.
5. "A"-"Not A" test	As for No. 4, but used where one of the samples has importance as a standard or reference product, is familiar to the subjects, or essential to the project as the current sample against which all other samples are measured.
6. Difference From Control test	Two samples which may show slight visual differences such as are caused by the normal heterogeneity of meats, vegetables, salads, and baked goods. Test is used where the size of the difference affects a decision about the test objective, e.g., in quality control and storage studies. Generally 30 to 50 presentations of the sample pair. Moderate amount of training required.
7. Sequential tests	Used with any of the above tests, to determine with a minimum of testing, at a predetermined significance level, whether the two samples are perceptibly (a) identical or (b) different.
8. Similarity testing	Used with any of the tests 1 to 6 when the test objective is to prove that no perceptible difference exists between two products. Used in situations such as (a) the substitution of a new ingredient for an old one that has become too expensive or unavailable or (b) a change in processing brought about by replacement of an old or inefficient piece of equipment.

Table 3
AREA OF APPLICATION OF ATTRIBUTE DIFFERENCE TESTS; HOW DOES ATTRIBUTE X DIFFER BETWEEN SAMPLES?

The tests in this table are used to determine whether or not, or the degree to which, two or more samples differ *with respect to one defined attribute*. This may be a single attribute such as sweetness, or a combination of several related attributes, such as freshness, or an overall evaluation, such as preference. With the exception of preference, panelists must be carefully trained to recognize the selected attribute, and the results are valid only to the extent that panelists understand and obey such instructions. A lack of difference in the selected attribute does *not* imply that no overall difference exists. Samples need not be visibly identical, as only the selected attribute is evaluated.

Test	Areas of Application
1. Paired Comparison test	One of the most-used attribute difference tests. Used to show which of two samples has more of the attribute under test (''Directional Difference test''), or which of two samples is preferred (''Paired Preference test''). Test exists in one- or two-sided applications. Generally 30 or more subjects, can be used with as few as 15.
2. Pairwise Ranking test	Used to rank three to six samples according to intensity of one attribute. Paired ranking is simple to perform and the statistical analysis is uncomplicated, but results are not as actionable as those obtained with rating. Generally 20 or more subjects, can be used with as few as 10.
3. Scheffé Paired Comparisons test	Used to rate two to six samples on a scale of intensity of the attribute under test. Samples may show slight differences in the attribute, such as are caused by the normal heterogeneity of meats, vegetables, salads, and baked goods. Test is used where the size of the difference affects a decision about the test objective, e.g., in quality control and in storage studies. Generally 14 or more subjects can be used with as few as 7.
4. Multiple Paired Comparison — Thurstone/Bradley/Morrissey	As No. 3 but used in special cases only, e.g., if testing for sample-by-panelist interaction is required (Thurstone/Bradley) and/or if not all pairs were presented (Morrissey). Generally 14 or more subjects can be used with as few as 7.
5. Simple Ranking test	Used to rank three to six, certainly no more than eight, samples according to one attribute. Ranking is simple to perform, but results are not as actionable as those obtained by rating: two samples of small or large difference in the attribute will show the same difference in rank (i.e., one rank unit). Ranking is useful to presort or screen samples for more detailed tests. Generally 16 or more subjects, can be used with as few as 8.
6. Rating of Several Samples	Used to rate three to six, certainly no more than eight samples on a numerical intensity scale according to one attribute. Similar to No. 3 except that it is a requirement that all samples be compared in one large set. Generally 16 or more subjects, can be used with as few as 8. May be used to compare descriptive analyses of several samples, but note (Chapter 6, Section III.F) that there will be some carryover (halo effect) between the attributes.
7. Balanced Incomplete Block test	As No. 5, but used when there are too many samples (e.g., 7 to 15) to be presented together in one sitting.
8. Rating of Several Samples, Balanced Incomplete Block	As No. 6, but used when there are too many samples (e.g., 7 to 15) to be presented together in one sitting.

Table 4
AREA OF APPLICATION OF AFFECTIVE TESTS USED IN CONSUMER TESTS AND EMPLOYEE ACCEPTANCE TESTS

Affective tests can be divided into *Preference tests* in which the task is to arrange the products tested in order of preference, *Acceptance tests* in which the task is to rate the product or products on a scale of acceptability, and *"attribute diagnostics"* in which the task is to rank or rate the principal attributes which determine a product's preference or acceptance. With regard to the statistical analysis, preference and acceptance tests can be seen as a special case of Attribute difference tests (Table 3), in which the attribute of interest is either preference or degree of acceptance. In theory, all tests listed in Table 3 can be used as Preference tests and/or as Acceptance tests. In practice, subjects in Affective tests are less experienced, and complex designs such as balanced incomplete blocks are not usable. The tests in this table are equally suitable for presentation in laboratory tests, employee acceptance tests, central location consumer tests, or home use consumer tests unless otherwise indicated.

Test	Question typically asked	Areas of application
Preference Tests		
1. Paired Preference	Which sample do you prefer? Which sample do you like better?	Comparison of two products
2. Rank Preference	Rank samples according to your preference with 1 = best, 2 = next best, etc.	Comparison of three to six products
3. Multiple Paired Preference	As No. 1.	Comparison of three to six products
4. Multiple Paired Preference, Selected Pairs	As No. 1.	Comparison of five to eight products
Acceptance Tests		
5. Simple Acceptance test	Is the sample acceptable/not acceptable?	First screening in Employee Acceptance test
6. Hedonic Rating	Chapter 9, Tables 2 to 5	One or more products to study how acceptance is distributed in the population represented by the subjects
Attribute Diagnostics		
7. Attribute-by-Preference test	Which sample did you prefer for fragrance?	Comparison of two to six products to determine which attributes "drive" preference
8. Hedonic Rating of Individual Attributes	Rate the following attributes on the hedonic scale provided	Study of one or more products, to determine which attributes, and at what level, "drive" preference
9. Intensity Rating of Individual Attributes	Rate the following attributes on the intensity scale provided, comparing with your ideal rating	Study of one or more products, in cases where groups of subjects differ in their preference

Table 5
AREA OF APPLICATION OF DESCRIPTIVE TESTS

Descriptive tests are very diverse, often designed or modified for each individual application, and therefore difficult to classify in a table such as this. A classification by inventor is perhaps the most helpful.

Test	Areas of application
1. Flavor Profile (Arthur D. Little)	In situations where many and varied samples must be judged by a few highly trained tasters
2. Texture Profile (General Foods)	In situations where many and varied samples must be judged for texture by a few highly trained tasters
3. QDA Method (Tragon Corp.)	In situations such as Quality Assurance in a large company, where large numbers of the same kind of products must be judged day in and day out by a well trained panel
4. Time-Intensity Descriptive Analysis	Useful for samples in which the perceived intensity of flavor varies over time after the product is taken into the mouth, e.g., bitterness of beer, sweetness of artificial sweeteners
5. Spectrum Method	A custom-design system suitable for most applications, including those under 1, 2, and 3 above
6. Modified, Short-Version Descriptive Analysis	To monitor a few critical attributes of a product through shelf-life studies; to examine possible manufacturing defects and product complaints; for routine Quality Assurance

REFERENCE

1. Sensory Evaluation Division, Institute of Food Technologists, Guidelines for the preparation and review of papers reporting sensory evaluation data, *Food Technol.*, 35(11), 50, 1981.

Chapter 14

GUIDELINES FOR REPORTING RESULTS

I. INTRODUCTION

For the user of sensory results, the most important consideration is how much confidence he can place in them. Two main factors determine this:[1]

1. Reliability: Would similar results be obtained if the test were repeated with the same panelists? With different panelists?
2. Validity: How valid are the conclusions? Did the test measure what it was intended to measure?

Because of the many opportunities for variability and bias resulting from the use of human subjects, reports of sensory tests must contain more detail than reports of physical or chemical measurements. It can be difficult to decide how much information to include; the recommendations below are mainly those of Prell[2] and the Sensory Evaluation Division of the Institute of Food Technologists.[3] Application of the suggested guidelines is illustrated in the Example at the end of the chapter.

II. SUMMARY

What did the test teach us? It is an important courtesy to the user not to oblige him or her to hunt through pages of text in order to discover the essence of the results. The conclusion is obvious to the sensory analyst and he should state it briefly and concisely in the opening summary. The summary should not exceed 110 words[2] and should answer the four what's:

- What was the objective?
- What was done?
- What were the results?
- What can be concluded?

III. OBJECTIVE

As reiterated many times in this book, a clear written formulation of the project objective and the test objective is fundamental to the success of any sensory experiment. The report (if directed to the project leader) should state and explain the test objective; if the report covers a complete project, it should state and explain the project objective as well as the objective of each test that formed part of the project.

In some cases, for example if the report is for publication, the explanation should take the form of an Introduction which includes a review, with references, of pertinent previous work. This should be followed by a brief definition of the problem. It is of great importance to state the approach which was taken to solve the problem; Table 1, Chapter 13, which follows the IFT,[3] should assist in this regard. If the study is based on a hypothesis, this hypothesis should be made evident to the reader in the introduction. Subsequent sections of the report should provide the test of the hypothesis.

Table 1
EXAMPLE OF REPORT: HOP CHARACTER IN FIVE BEERS

Summary

What was the objective?
What was done?

What were the results?
What can be concluded?

In order to choose among five lots of hops on the basis of the amount of hop character they are likely to provide, pilot brews were made with hop samples 1,2,3,4, and 5 costing $1.00, $1.20, $1.40, $1.60 and $1.80/lb; 20 trained members of the brewery panel judged each beer three times on a scale from 0 to 9. Sample 4 received a rating of 3.9, significantly higher than samples 2 and 5, at 3.0 and 2.9. Samples 1 and 3 were significantly lowest at 2.1 and 1.4. It can be concluded that hop samples 4 and 2 deliver more hop character per dollar than the remainder.

Objectives

Project objective,
  test objectives,
  agreed before the
  experiment

The brewery obtained representative lot samples from five suppliers. The project objective was to choose among the lots based on their ability to provide hop character and the test objectives, (1) to compare the five beers for degree of hop character on a meaningful scale and (2) to obtain a measure of the reliability of the results.

Experimental

Design which accom-
  plishes objectives 1 and 2

Describe sensory tests used

Describe panel: number,
  training, etc.
Describe conditions of test:
  Screening of samples

  Information to panel
  Panel area
  Sample presentation

Statistical techniques

Design — The five samples were test brewed to produce a standard bitterness level of 14 BU. The test beers were evaluated by 20 selected members of the brewery panel. The test set was tasted three times on separate days.
Sensory evaluation — The tasters evaluated the amount of hop character on a scale of 0 to 9. Reference standards were available as follows: synthetic hop character at 1.0 mg/ℓ = 3.0 scale units, and at 3 mg/ℓ = 6.0 scale units.
The Panel — 20 panel members were selected on the basis of past performance evaluating hop character. All 20 panelists tasted all three sets.
Sample preparation and presentation — The test beers were stored at 12°C and tasted 7—10 days after bottling. Samples were screened by two experienced tasters who found them representative of the type of beer with no differences in color, foam, or flavor other than hop character. Panel members were informed that samples were test brews with different hops, but the identity of individual samples was not disclosed. Members worked individually in booths and no discussion took place after tasting. Sample portions of 70 mℓ were served at 12°C in clear 8-oz glasses. The five samples were presented simultaneously in balanced, random order. Samples were swallowed.
Statistical evaluation — Results were evaluated by split-plot analysis of variance.

Results and Discussion

Present results concisely

The average results for the five beers are shown in Table I and the corresponding statistical analysis in Table II. Sample 4 received a significantly higher rating for hop character (3.9) than the remaining samples.

Give enough data to justify
  conclusions

Table I
AVERAGE HOP CHARACTER RATINGS
FOR THE FIVE BEER SAMPLES

Sample	4	2	5	1	3
Mean	3.9ᵃ	3.0	2.9	2.1	1.4
	0.36	0.38	0.34	0.32	0.35

[a]   (Samples not connected by a common underscore are significantly different at the 5% significance level)

Table 1 (continued)
EXAMPLE OF REPORT: HOP CHARACTER IN FIVE BEERS

Give probability levels, degrees of freedom, obtained value of test statistic

Table II
SPLIT-PLOT ANOVA OF THE RESULTS

Source of variation	Degrees of freedom	Sum of squares	Mean squares	F
Total	299	976.25		
Replications	2	7.94		
Subjects	19	412.78	21.73	33.94**
Error(A)	38	24.33	0.64	
Samples	4	209.40	52.35	38.83**
Subjects X Samples	76	106.07	1.40	1.04
Error(B)	160	215.73	1.35	

** = significant at the 1% level.

Interpret the data, following the design of the experiment

Samples 2 and 5, with nearly identical ratings of 3.0 and 2.9 had significantly less hop character than sample 4, but significantly more than samples 1 and 3. Sample 3 had significantly less hop character than all other samples. The statistical evaluation shows no significance for the subject-by-sample interaction (F = 1.04), so it may be assumed that the panelists were consistent in their ratings. The significance of the subject effect (F = 33.94) suggests that the panelists used different parts of the scale to express their perceptions; this is not uncommon. Further, when there is no interaction, the subject-to-subject differences are of secondary interest. The primary concern, the difference among samples, was evaluated using an HSD multiple comparison procedure; $HSD_{5\%} = 0.6$ which results in the differences shown by underscoring in Table I. Variations in the amounts of hops used to obtain the BU level of 14 were small compared with the variations in perceived hop character intensity.

Conclusions

End with clear-cut conclusions

Of the five samples tested, sample 4 ($1.60/lb) produced a significantly higher level of hop character. Sample 2 ($1.20) merits consideration for less expensive beers.

IV. EXPERIMENTAL

The experimental section should provide sufficient detail to allow the work to be repeated. Accepted methods should be cited by adequate references. It is sometimes overlooked that subheadings in the experimental section help the reader find specific information. The section should describe the important steps in collecting the sensory data and will usually include the following.

Experimental design — Assuming that the objective was clearly stated above, the text should now explain the "layout" of the experiment in terms of the objective. If there are major and minor objectives, the report should show how this is reflected in the design. If an advanced design is used (randomized complete block, balanced incomplete block, Latin square, etc.) it can be described by reference to the appropriate section of Cochran and Cox.[4] Next, state the measurements made (e.g., sensory, physical, chemical), sample variables and level of the variables (where appropriate), number of replications, and limitations of the design (e.g., lots available for sampling, nature, and number of samples evaluated in a test session). Describe the efforts made to reduce the experimental error.

Sensory methods — Refer to the methods employed by the terminology used in this book, see Tables 2 to 5 in Chapter 13, which are the same as that of the International Standards Organization[5] and the IFT.[3]

The panel — The number of panelists for each experimental condition should be stated as it influences the statistical significance of the results obtained. If too few panelists are used, large differences are required for statistical significance, whereas if too many are used, e.g., 1000 for a triangular test, statistical significance may result when the actual difference is too small to have practical meaning. Changes in the panel during the course of the experiments should be avoided, but if they do occur, they must be fully described. The extent of previous training and the methods used to prepare the panelists for the current tests, including full description of any reference standards used, are important information needed to judge the validity of the results. The composition of the panel (age, sex, etc.) should be described if any affective tests were part of the experiment.

Conditions of the test — The physical conditions of the test area as well as the way samples are prepared and presented are important variables which influence both reliability and validity of the results. The report should contain the following information:

1. Test area — The location of the test area (booth, store, home, bus) should be stated, and any distractions present (odors, noise, heat, cold, lighting) described together with efforts made to minimize their influence.
2. Sample preparation — The equipment and methods of sample preparation should be described (time, temperature, any carrier used). Identify and describe raw materials and formulations if applicable.
3. Sample presentation — The description should enable the reader to judge the degree of bias likely to be contained in the results and may include any of the following capable of influencing them:

- Did panelists work individually or as a group
- Lighting used if different from normal
- Sample quantity, container, utensils, temperature
- Order of presentation (randomized, balanced)
- Coding of sample containers, e.g., three-digit random numbers
- Any special instructions such as mouth rinsing, information about the identity of samples or variable under test; time intervals between samples; were samples swallowed or expectorated
- Any other variable which could influence the results, e.g., time of day, high or low humidity, age of samples, etc., etc.

Statistical techniques — The manner in which the data reported were derived from actual test responses should be defined, e.g., conversion of scores to ranks. The type of statistical analysis used and the degree to which underlying assumptions (e.g., normality) are met should be discussed, as should the null hypothesis and alternate hypothesis, if not trivial.

V. RESULTS AND DISCUSSION

Results should be presented concisely in the form of tables and figures, and enough data should be given to justify conclusions. However, the same information should not be presented in both forms. Tabular data generally are more concise except for trends and interactions which may be easier to see from figures.

The results section should summarize the relevant collected data and the statistical analyses. All results should be shown, including those which run counter to the hypotheses. Reports of tests of significance (F, χ^2, t, r, etc.) should list the probability level, the degrees of freedom if applicable, the obtained value of the test statistic, and the direction of the effect.

In the discussion section, the theoretical and practical significance of the results should be pointed out and related to previous knowledge. The discussion should begin by briefly stating whether the results support or fail to support any original hypothesis. The interpretation of data should be logically organized and should follow the design of the experiment. The results should be interpreted, compared, and contrasted (with limitations indicated), and the report should end with clear-cut conclusions.

See Table 1 and Reference 1 which illustrate the development of terminology and scales for a descriptive study.

REFERENCES

1. **Larmond, E.,** Better reports of sensory evaluation, *Tech. Q. Master Brewers Assoc. Am.*, 18, 7, 1981.
2. **Prell, P. A.,** Preparation of reports and manuscripts which include sensory evaluation data, *Food Technol.*, 30(11), 40, 1976.
3. Sensory Evaluation Division, Institute of Food Technologists, Guidelines for the preparation and review of papers reporting sensory evaluation data, *Food Technol.*, 35(4), 16; 35(11), 50, 1981.
4. **Cochran, W. G. and Cox, G. M.,** *Experimental Designs*, John Wiley & Sons, New York, 1957, 469.
5. International Standards Organization, *International Standard ISO 6658, Sensory Analysis — Methodology — General Guidance*, ISO, Tour Europe, Paris, 1983.

STATISTICAL TABLES

Table T1
RANDOM ORDERS OF THE DIGITS 1 TO 9. ARRANGED IN GROUPS OF THREE COLUMNS

Instructions

(1) To generate a sequence of three-digit random numbers, enter the table at any location, e.g., closing the eyes and pointing. Without inspecting the numbers, decide whether to move up or down the column entered. Record as many numbers as needed. Discard any numbers that are unsuitable (out of range, came up before, etc.). The sequence of numbers obtained in this manner is in random order.

(2) To generate a sequence of two-digit random numbers, proceed as above, but first decide, e.g., by coin toss, whether to use the first two or last two digits of each number taken from the table. Treat each three-digit number in the same manner, i.e., discard the same digit from each. If a two-digit number comes up more than once, retain only the first.

(3) Random number tables are impractical for problems such as: "place the numbers from 15 to 50 in random order". Instead, write each number on a card and draw the cards blindly from a bag or use a computerized random number generator such as PROC PLAN from SAS.®

```
862 245 458 396 522 498 298 665 635 665 113 917 365 332 896 314 688 468 663 712 585 351 847
223 398 183 765 138 369 163 743 593 252 581 355 542 691 537 222 746 636 478 368 949 797 295
756 954 266 174 496 133 759 488 854 187 228 824 881 549 759 169 122 919 946 293 874 289 452
544 537 522 459 984 585 946 127 711 549 445 793 734 855 121 885 595 152 237 574 611 145 784
681 829 614 547 869 742 822 554 448 813 976 688 959 714 912 646 873 397 159 155 136 463 363
199 113 941 933 375 651 414 891 129 938 862 572 698 128 363 478 214 841 314 437 792 874 926
918 481 797 621 743 827 377 916 966 426 657 246 423 277 685 533 937 223 582 946 323 626 519
335 662 875 282 617 274 635 379 287 791 334 139 117 963 448 957 451 585 821 829 267 512 638
477 776 339 818 251 916 581 232 372 374 799 461 276 486 274 791 369 774 795 681 458 938 171

653 489 538 216 446 849 914 337 993 459 325 614 771 244 429 874 557 119 122 417 882 714 769
749 824 721 967 287 556 628 843 725 731 553 253 183 653 988 431 788 426 875 838 457 927 475
522 967 259 532 618 624 396 562 134 563 932 441 834 787 231 958 232 537 439 956 531 345 352
475 172 986 859 925 932 282 924 842 642 797 565 399 896 596 282 441 784 258 684 625 662 291
894 333 612 728 869 487 741 259 476 127 286 736 257 168 847 316 969 692 786 549 949 559 526
116 218 464 191 132 218 573 786 258 296 471 372 618 935 353 747 123 863 644 161 793 196 847
381 641 393 375 354 193 165 615 587 384 119 187 965 572 112 695 615 941 361 375 376 871 633
968 755 847 643 773 765 439 478 611 978 868 898 546 319 775 169 896 275 513 222 114 233 184

742 421 226 286 522 618 471 218 397 745 461 477 478 535 957 674 132 228 442 225 444 171 151
859 878 392 311 659 772 935 447 834 117 658 161 754 654 176 883 855 195 637 751 586 948 513
964 593 137 574 288 994 582 961 746 336 983 782 611 988 833 265 969 584 564 683 197 214 326
177 636 674 897 167 157 856 524 662 598 145 926 362 777 415 931 313 317 195 137 959 536 985
228 755 915 955 946 233 647 653 425 674 719 543 549 826 669 429 576 773 756 392 632 725 879
591 214 851 669 394 349 299 192 179 261 332 294 896 299 782 397 791 659 921 569 811 683 762
636 167 789 438 413 565 118 889 253 452 577 859 125 141 241 746 444 841 313 446 225 362 248
415 982 543 743 835 826 364 776 988 923 224 615 283 462 328 512 228 466 278 874 373 499 437
383 349 468 122 771 481 723 335 511 889 896 338 937 313 594 158 687 932 889 918 768 857 694

975 973 235 811 761 226 637 382 741 767 894 371 128 972 161 911 427 164 461 991 792 256 194
257 752 667 227 813 488 598 198 979 388 921 926 715 349 644 846 879 242 695 222 633 595 526
723 395 174 453 276 732 323 866 583 826 562 817 397 556 786 358 755 996 249 676 461 614 485
448 524 951 982 455 999 451 434 695 693 788 493 951 231 259 667 318 655 374 559 577 873 747
539 881 529 664 594 555 779 629 168 442 377 685 449 128 532 232 241 418 536 733 348 162 919
661 469 312 748 942 671 284 777 354 939 116 158 583 615 977 525 193 871 883 818 154 449 333
394 647 493 599 628 317 846 255 416 174 449 269 276 883 828 193 984 529 758 164 215 938 272
882 216 786 376 187 864 912 941 837 551 233 744 634 464 313 474 536 333 927 345 889 387 658
116 138 848 135 339 143 165 513 222 215 655 532 862 797 495 789 662 787 112 487 926 721 861
```

Table T1 (continued)
RANDOM ORDERS OF THE DIGITS 1 TO 9. ARRANGED IN GROUPS OF THREE COLUMNS

```
256 654 283 248 626 424 139 819 747 496 134 143 741 552 698 876 441 796 671 833 337 168 952
811 576 571 877 278 311 683 726 585 672 419 597 896 221 365 962 227 145 993 211 275 656 294
492 393 666 634 347 582 358 167 132 824 328 732 464 945 486 633 169 511 129 765 193 485 511
784 819 449 356 835 193 966 283 253 169 547 654 653 619 533 327 672 434 816 149 768 939 328
545 722 394 599 482 935 491 534 464 948 253 411 277 134 951 759 715 677 554 627 459 792 187
323 961 817 765 169 756 247 678 811 583 671 365 989 773 117 591 853 388 268 954 881 241 849
169 235 732 923 711 677 715 355 698 331 766 986 532 397 772 484 386 253 737 478 626 577 666
637 188 928 411 993 249 572 492 926 217 985 279 118 488 829 245 994 822 382 396 514 324 733
978 447 155 182 554 868 824 941 379 755 892 818 325 866 244 118 538 969 445 582 942 813 475
```

From Cochran, W. G. and Cox, G. M., *Experimental Design*, John Wiley & Sons, New York, 1957. With permission.

Table T2
BINOMIAL PROBABILITIES[a]

Instructions: See the Examples in Chapter 11

n	x	0.05	0.10	0.15	0.20	0.25	0.30	0.35	0.40	0.45	0.50
1	0	0.9500	0.9000	0.8500	0.8000	0.7500	0.7000	0.6500	0.6000	0.5500	0.5000
	1	0.0500	0.1000	0.1500	0.2000	0.2500	0.3000	0.3500	0.4000	0.4500	0.5000
2	0	0.9025	0.8100	0.7225	0.6400	0.5625	0.4900	0.4225	0.3600	0.3025	0.2500
	1	0.0950	0.1800	0.2550	0.3200	0.3750	0.4200	0.4550	0.4800	0.4950	0.5000
	2	0.0025	0.0100	0.0225	0.0400	0.0625	0.0900	0.1225	0.1600	0.2025	0.2500
3	0	0.8574	0.7290	0.6141	0.5120	0.4219	0.3430	0.2746	0.2160	0.1664	0.1250
	1	0.1354	0.2430	0.3251	0.3840	0.4219	0.4410	0.4436	0.4320	0.4084	0.3750
	2	0.0071	0.0270	0.0574	0.0960	0.1406	0.1890	0.2389	0.2880	0.3341	0.3750
	3	0.0001	0.0010	0.0034	0.0080	0.0156	0.0270	0.0429	0.0640	0.0911	0.1250
4	0	0.8145	0.6561	0.5220	0.4096	0.3164	0.2401	0.1785	0.1296	0.0915	0.0625
	1	0.1715	0.2916	0.3685	0.4096	0.4219	0.4116	0.3845	0.3456	0.2995	0.2500
	2	0.0135	0.0486	0.0975	0.1536	0.2109	0.2646	0.3105	0.3456	0.3675	0.3750
	3	0.0005	0.0036	0.0115	0.0256	0.0469	0.0756	0.1115	0.1536	0.2005	0.2500
	4	0.0000	0.0001	0.0005	0.0016	0.0039	0.0081	0.0150	0.0256	0.0410	0.0625
5	0	0.7738	0.5905	0.4437	0.3277	0.2373	0.1681	0.1160	0.0778	0.0503	0.0312
	1	0.2036	0.3280	0.3915	0.4096	0.3955	0.3602	0.3124	0.2592	0.2059	0.1562
	2	0.0214	0.0729	0.1382	0.2048	0.2637	0.3087	0.3364	0.3456	0.3369	0.3125
	3	0.0011	0.0081	0.0244	0.0512	0.0879	0.1323	0.1811	0.2304	0.2757	0.3125
	4	0.0000	0.0004	0.0022	0.0064	0.0146	0.0284	0.0488	0.0768	0.1128	0.1562
	5	0.0000	0.0000	0.0001	0.0003	0.0010	0.0024	0.0053	0.0102	0.0185	0.0312
6	0	0.7351	0.5314	0.3771	0.2621	0.1780	0.1176	0.0754	0.0467	0.0277	0.0156
	1	0.2321	0.3543	0.3993	0.3932	0.3560	0.3025	0.2437	0.1866	0.1359	0.0938
	2	0.0305	0.0984	0.1762	0.2458	0.2966	0.3241	0.3280	0.3110	0.2780	0.2344
	3	0.0021	0.0146	0.0415	0.0819	0.1318	0.1852	0.2355	0.2765	0.3032	0.3125
	4	0.0001	0.0012	0.0055	0.0154	0.0330	0.0595	0.0951	0.1382	0.1861	0.2344
	5	0.0000	0.0001	0.0004	0.0015	0.0044	0.0102	0.0205	0.0369	0.0609	0.0938
	6	0.0000	0.0000	0.0000	0.0001	0.0002	0.0007	0.0018	0.0041	0.0083	0.0156

Table T2 (continued)
BINOMIAL PROBABILITIES[a]

Instructions: See the Examples in Chapter 11

p

n	x	0.05	0.10	0.15	0.20	0.25	0.30	0.35	0.40	0.45	0.50
7	0	0.6983	0.4783	0.3206	0.2097	0.1335	0.0824	0.0490	0.0280	0.0152	0.0078
	1	0.2573	0.3720	0.3960	0.3670	0.3115	0.2471	0.1848	0.1306	0.0872	0.0547
	2	0.0406	0.1240	0.2097	0.2753	0.3115	0.3177	0.2985	0.2613	0.2140	0.1641
	3	0.0036	0.0230	0.0617	0.1147	0.1730	0.2269	0.2679	0.2903	0.2918	0.2734
	4	0.0002	0.0026	0.0109	0.0287	0.0577	0.0972	0.1442	0.1935	0.2388	0.2734
	5	0.0000	0.0002	0.0012	0.0043	0.0115	0.0250	0.0466	0.0774	0.1172	0.1641
	6	0.0000	0.0000	0.0001	0.0004	0.0013	0.0036	0.0084	0.0172	0.0320	0.0547
	7	0.0000	0.0000	0.0000	0.0000	0.0001	0.0002	0.0006	0.0016	0.0037	0.0078
8	0	0.6634	0.4305	0.2725	0.1678	0.1001	0.0576	0.0319	0.0168	0.0084	0.0039
	1	0.2793	0.3826	0.3847	0.3355	0.2670	0.1977	0.1373	0.0896	0.0548	0.0312
	2	0.0515	0.1488	0.2376	0.2936	0.3115	0.2965	0.2587	0.2090	0.1569	0.1094
	3	0.0054	0.0331	0.0839	0.1468	0.2076	0.2541	0.2786	0.2787	0.2568	0.2188
	4	0.0004	0.0046	0.0185	0.0459	0.0865	0.1361	0.1875	0.2322	0.2627	0.2734
	5	0.0000	0.0004	0.0026	0.0092	0.0231	0.0467	0.0808	0.1239	0.1719	0.2188
	6	0.0000	0.0000	0.0002	0.0011	0.0038	0.0100	0.0217	0.0413	0.0703	0.1094
	7	0.0000	0.0000	0.0000	0.0001	0.0004	0.0012	0.0033	0.0079	0.0164	0.0312
	8	0.0000	0.0000	0.0000	0.0000	0.0000	0.0001	0.0002	0.0007	0.0017	0.0039
9	0	0.6302	0.3874	0.2316	0.1342	0.0751	0.0404	0.0207	0.0101	0.0046	0.0020
	1	0.2985	0.3874	0.3679	0.3020	0.2253	0.1556	0.1004	0.0605	0.0339	0.0176
	2	0.0629	0.1722	0.2597	0.3020	0.3003	0.2668	0.2162	0.1612	0.1110	0.0703
	3	0.0077	0.0446	0.1069	0.1762	0.2336	0.2668	0.2716	0.2508	0.2119	0.1641
	4	0.0006	0.0074	0.0283	0.0661	0.1168	0.1715	0.2194	0.2508	0.2600	0.2461
	5	0.0000	0.0008	0.0050	0.0165	0.0389	0.0735	0.1181	0.1672	0.2128	0.2461
	6	0.0000	0.0001	0.0006	0.0028	0.0087	0.0210	0.0424	0.0743	0.1160	0.1641
	7	0.0000	0.0000	0.0000	0.0003	0.0012	0.0039	0.0098	0.0212	0.0407	0.0703
	8	0.0000	0.0000	0.0000	0.0000	0.0001	0.0004	0.0013	0.0035	0.0083	0.0176
	9	0.0000	0.0000	0.0000	0.0000	0.0000	0.0000	0.0001	0.0003	0.0008	0.0020
10	0	0.5987	0.3487	0.1969	0.1074	0.0563	0.0282	0.0135	0.0060	0.0025	0.0010
	1	0.3151	0.3874	0.3474	0.2684	0.1877	0.1211	0.0725	0.0403	0.0207	0.0098
	2	0.0746	0.1937	0.2759	0.3020	0.2816	0.2335	0.1757	0.1209	0.0763	0.0439
	3	0.0105	0.0574	0.1298	0.2013	0.2503	0.2668	0.2522	0.2150	0.1665	0.1172
	4	0.0010	0.0112	0.0401	0.0881	0.1460	0.2001	0.2377	0.2508	0.2384	0.2051
	5	0.0001	0.0015	0.0085	0.0264	0.0584	0.1029	0.1536	0.2007	0.2340	0.2461
	6	0.0000	0.0001	0.0012	0.0055	0.0162	0.0368	0.0689	0.1115	0.1596	0.2051
	7	0.0000	0.0000	0.0001	0.0008	0.0031	0.0090	0.0212	0.0425	0.0746	0.1172
	8	0.0000	0.0000	0.0000	0.0001	0.0004	0.0014	0.0043	0.0106	0.0229	0.0439
	9	0.0000	0.0000	0.0000	0.0000	0.0000	0.0001	0.0005	0.0016	0.0042	0.0098
	10	0.0000	0.0000	0.0000	0.0000	0.0000	0.0000	0.0000	0.0001	0.0003	0.0010
11	0	0.5688	0.3138	0.1673	0.0859	0.0422	0.0198	0.0088	0.0036	0.0014	0.0004
	1	0.3293	0.3835	0.3248	0.2362	0.1549	0.0932	0.0518	0.0266	0.0125	0.0055
	2	0.0867	0.2131	0.2866	0.2953	0.2581	0.1998	0.1395	0.0887	0.0513	0.0269
	3	0.0137	0.0710	0.1517	0.2215	0.2581	0.2568	0.2254	0.1774	0.1259	0.0806
	4	0.0014	0.0158	0.0536	0.1107	0.1721	0.2201	0.2428	0.2365	0.2060	0.1611

Table T2 (continued)
BINOMIAL PROBABILITIES[a]

Instructions: See the Examples in Chapter 11

p

n	x	0.05	0.10	0.15	0.20	0.25	0.30	0.35	0.40	0.45	0.50
	5	0.0001	0.0025	0.0132	0.0388	0.0803	0.1321	0.1830	0.2207	0.2360	0.2256
	6	0.0000	0.0003	0.0023	0.0097	0.0268	0.0566	0.0985	0.1471	0.1931	0.2256
	7	0.0000	0.0000	0.0003	0.0017	0.0064	0.0173	0.0379	0.0701	0.1128	0.1611
	8	0.0000	0.0000	0.0000	0.0002	0.0011	0.0037	0.0102	0.0234	0.0462	0.0806
	9	0.0000	0.0000	0.0000	0.0000	0.0001	0.0005	0.0018	0.0052	0.0126	0.0269
	10	0.0000	0.0000	0.0000	0.0000	0.0000	0.0000	0.0002	0.0007	0.0021	0.0054
	11	0.0000	0.0000	0.0000	0.0000	0.0000	0.0000	0.0000	0.0000	0.0002	0.0005
12	0	0.5404	0.2824	0.1422	0.0687	0.0317	0.0138	0.0057	0.0022	0.0008	0.0002
	1	0.3413	0.3766	0.3012	0.2062	0.1267	0.0712	0.0368	0.0174	0.0075	0.0029
	2	0.0988	0.2301	0.2924	0.2835	0.2323	0.1678	0.1088	0.0639	0.0339	0.0161
	3	0.0173	0.0852	0.1720	0.2362	0.2581	0.2397	0.1954	0.1419	0.0923	0.0537
	4	0.0021	0.0213	0.0683	0.1329	0.1936	0.2311	0.2367	0.2128	0.1700	0.1208
	5	0.0002	0.0038	0.0193	0.0532	0.1032	0.1585	0.2039	0.2270	0.2225	0.1934
	6	0.0000	0.0005	0.0040	0.0155	0.0401	0.0792	0.1281	0.1766	0.2124	0.2256
	7	0.0000	0.0000	0.0006	0.0033	0.0115	0.0291	0.0591	0.1009	0.1489	0.1934
	8	0.0000	0.0000	0.0001	0.0005	0.0024	0.0078	0.0199	0.0420	0.0762	0.1208
	9	0.0000	0.0000	0.0000	0.0001	0.0004	0.0015	0.0048	0.0125	0.0277	0.0537
	10	0.0000	0.0000	0.0000	0.0000	0.0000	0.0002	0.0008	0.0025	0.0068	0.0161
	11	0.0000	0.0000	0.0000	0.0000	0.0000	0.0000	0.0001	0.0003	0.0010	0.0029
	12	0.0000	0.0000	0.0000	0.0000	0.0000	0.0000	0.0000	0.0000	0.0001	0.0002
13	0	0.5133	0.2542	0.1209	0.0550	0.0238	0.0097	0.0037	0.0013	0.0004	0.0001
	1	0.3512	0.3672	0.2774	0.1787	0.1029	0.0540	0.0259	0.0113	0.0045	0.0016
	2	0.1109	0.2448	0.2937	0.2680	0.2059	0.1388	0.0836	0.0453	0.0220	0.0095
	3	0.0214	0.0997	0.1900	0.2457	0.2517	0.2181	0.1651	0.1107	0.0660	0.0349
	4	0.0028	0.0277	0.0838	0.1535	0.2097	0.2337	0.2222	0.1845	0.1350	0.0873
	5	0.0003	0.0055	0.0266	0.0691	0.1258	0.1803	0.2154	0.2214	0.1989	0.1571
	6	0.0000	0.0008	0.0063	0.0230	0.0559	0.1030	0.1546	0.1968	0.2169	0.2095
	7	0.0000	0.0001	0.0011	0.0058	0.0186	0.0442	0.0833	0.1312	0.1775	0.2095
	8	0.0000	0.0000	0.0001	0.0011	0.0047	0.0142	0.0336	0.0656	0.1089	0.1571
	9	0.0000	0.0000	0.0000	0.0001	0.0009	0.0034	0.0101	0.0243	0.0495	0.0873
	10	0.0000	0.0000	0.0000	0.0000	0.0001	0.0006	0.0022	0.0065	0.0162	0.0349
	11	0.0000	0.0000	0.0000	0.0000	0.0000	0.0001	0.0003	0.0012	0.0036	0.0095
	12	0.0000	0.0000	0.0000	0.0000	0.0000	0.0000	0.0000	0.0001	0.0005	0.0016
	13	0.0000	0.0000	0.0000	0.0000	0.0000	0.0000	0.0000	0.0000	0.0000	0.0001
14	0	0.4877	0.2288	0.1028	0.0440	0.0178	0.0068	0.0024	0.0008	0.0002	0.0001
	1	0.3593	0.3559	0.2539	0.1539	0.0832	0.0407	0.0181	0.0073	0.0027	0.0009
	2	0.1229	0.2570	0.2912	0.2501	0.1802	0.1134	0.0634	0.0317	0.0141	0.0056
	3	0.0259	0.1142	0.2056	0.2501	0.2402	0.1943	0.1366	0.0845	0.0462	0.0222
	4	0.0037	0.0349	0.0998	0.1720	0.2202	0.2290	0.2022	0.1549	0.1040	0.0611
	5	0.0004	0.0078	0.0352	0.0860	0.1468	0.1963	0.2178	0.2066	0.1701	0.1222
	6	0.0000	0.0013	0.0093	0.0322	0.0734	0.1262	0.1759	0.2066	0.2088	0.1833
	7	0.0000	0.0002	0.0019	0.0092	0.0280	0.0618	0.1082	0.1574	0.1952	0.2095
	8	0.0000	0.0000	0.0003	0.0020	0.0082	0.0232	0.0510	0.0918	0.1398	0.1833
	9	0.0000	0.0000	0.0000	0.0003	0.0018	0.0066	0.0183	0.0408	0.0762	0.1222

Table T2 (continued)
BINOMIAL PROBABILITIES[a]

Instructions: See the Examples in Chapter 11

p

n	x	0.05	0.10	0.15	0.20	0.25	0.30	0.35	0.40	0.45	0.50
	10	0.0000	0.0000	0.0000	0.0000	0.0003	0.0014	0.0049	0.0136	0.0312	0.0611
	11	0.0000	0.0000	0.0000	0.0000	0.0000	0.0002	0.0010	0.0033	0.0093	0.0222
	12	0.0000	0.0000	0.0000	0.0000	0.0000	0.0000	0.0001	0.0005	0.0019	0.0056
	13	0.0000	0.0000	0.0000	0.0000	0.0000	0.0000	0.0000	0.0001	0.0002	0.0009
	14	0.0000	0.0000	0.0000	0.0000	0.0000	0.0000	0.0000	0.0000	0.0000	0.0001
15	0	0.4633	0.2059	0.0874	0.0352	0.0134	0.0047	0.0016	0.0005	0.0001	0.0000
	1	0.3658	0.3432	0.2312	0.1319	0.0668	0.0305	0.0126	0.0047	0.0016	0.0005
	2	0.1348	0.2669	0.2856	0.2309	0.1559	0.0916	0.0476	0.0219	0.0090	0.0032
	3	0.0307	0.1285	0.2184	0.2501	0.2252	0.1700	0.1110	0.0634	0.0318	0.0139
	4	0.0049	0.0428	0.1156	0.1876	0.2252	0.2186	0.1792	0.1268	0.0780	0.0417
	5	0.0006	0.0105	0.0449	0.1032	0.1651	0.2061	0.2123	0.1859	0.1404	0.0916
	6	0.0000	0.0019	0.0132	0.0430	0.0917	0.1472	0.1906	0.2066	0.1914	0.1527
	7	0.0000	0.0003	0.0030	0.0138	0.0393	0.0811	0.1319	0.1771	0.2013	0.1964
	8	0.0000	0.0000	0.0005	0.0035	0.0131	0.0348	0.0710	0.1181	0.1647	0.1964
	9	0.0000	0.0000	0.0001	0.0007	0.0034	0.0116	0.0298	0.0612	0.1048	0.1527
	10	0.0000	0.0000	0.0000	0.0001	0.0007	0.0030	0.0096	0.0245	0.0515	0.0916
	11	0.0000	0.0000	0.0000	0.0000	0.0001	0.0006	0.0024	0.0074	0.0191	0.0417
	12	0.0000	0.0000	0.0000	0.0000	0.0000	0.0001	0.0004	0.0016	0.0052	0.0139
	13	0.0000	0.0000	0.0000	0.0000	0.0000	0.0000	0.0001	0.0003	0.0010	0.0032
	14	0.0000	0.0000	0.0000	0.0000	0.0000	0.0000	0.0000	0.0000	0.0001	0.0005
	15	0.0000	0.0000	0.0000	0.0000	0.0000	0.0000	0.0000	0.0000	0.0000	0.0000
16	0	0.4401	0.1853	0.0743	0.0281	0.0100	0.0033	0.0010	0.0003	0.0001	0.0000
	1	0.3706	0.3294	0.2097	0.1126	0.0535	0.0228	0.0087	0.0030	0.0009	0.0002
	2	0.1463	0.2745	0.2775	0.2111	0.1336	0.0732	0.0353	0.0150	0.0056	0.0018
	3	0.0359	0.1423	0.2285	0.2463	0.2079	0.1465	0.0888	0.0468	0.0215	0.0085
	4	0.0061	0.0514	0.1311	0.2001	0.2252	0.2040	0.1553	0.1014	0.0572	0.0278
	5	0.0008	0.0137	0.0555	0.1201	0.1802	0.2099	0.2008	0.1623	0.1123	0.0667
	6	0.0001	0.0028	0.0180	0.0550	0.1101	0.1649	0.1982	0.1983	0.1684	0.1222
	7	0.0000	0.0004	0.0045	0.0197	0.0524	0.1010	0.1524	0.1889	0.1969	0.1746
	8	0.0000	0.0001	0.0009	0.0055	0.0197	0.0487	0.0923	0.1417	0.1812	0.1964
	9	0.0000	0.0000	0.0001	0.0012	0.0058	0.0185	0.0442	0.0840	0.1318	0.1746
	10	0.0000	0.0000	0.0000	0.0002	0.0014	0.0056	0.0167	0.0392	0.0755	0.1222
	11	0.0000	0.0000	0.0000	0.0000	0.0002	0.0013	0.0049	0.0142	0.0337	0.0667
	12	0.0000	0.0000	0.0000	0.0000	0.0000	0.0002	0.0011	0.0040	0.0115	0.0278
	13	0.0000	0.0000	0.0000	0.0000	0.0000	0.0000	0.0002	0.0008	0.0029	0.0085
	14	0.0000	0.0000	0.0000	0.0000	0.0000	0.0000	0.0000	0.0001	0.0005	0.0018
	15	0.0000	0.0000	0.0000	0.0000	0.0000	0.0000	0.0000	0.0000	0.0001	0.0002
	16	0.0000	0.0000	0.0000	0.0000	0.0000	0.0000	0.0000	0.0000	0.0000	0.0000
17	0	0.4181	0.1668	0.0631	0.0225	0.0075	0.0023	0.0007	0.0002	0.0000	0.0000
	1	0.3741	0.3150	0.1893	0.0957	0.0426	0.0169	0.0060	0.0019	0.0005	0.0001
	2	0.1575	0.2800	0.2673	0.1914	0.1136	0.0581	0.0260	0.0102	0.0035	0.0010
	3	0.0415	0.1556	0.2359	0.2393	0.1893	0.1245	0.0701	0.0341	0.0144	0.0052
	4	0.9076	0.0605	0.1457	0.2093	0.2209	0.1868	0.1320	0.0796	0.0411	0.0182

Table T2 (continued)
BINOMIAL PROBABILITIES[a]

Instructions: See the Examples in Chapter 11

p

n	x	0.05	0.10	0.15	0.20	0.25	0.30	0.35	0.40	0.45	0.50
	5	0.0010	0.0175	0.0668	0.1361	0.1914	0.2081	0.1849	0.1379	0.0875	0.0472
	6	0.0001	0.0039	0.0236	0.0680	0.1276	0.1784	0.1991	0.1839	0.1432	0.0944
	7	0.0000	0.0007	0.0065	0.0267	0.0668	0.1201	0.1685	0.1927	0.1841	0.1484
	8	0.0000	0.0001	0.0014	0.0084	0.0279	0.0644	0.1134	0.1606	0.1883	0.1855
	9	0.0000	0.0000	0.0003	0.0021	0.0093	0.0276	0.0611	0.1070	0.1540	0.1855
	10	0.0000	0.0000	0.0000	0.0004	0.0025	0.0095	0.0263	0.0571	0.1008	0.1484
	11	0.0000	0.0000	0.0000	0.0001	0.0005	0.0026	0.0090	0.0242	0.0525	0.0944
	12	0.0000	0.0000	0.0000	0.0000	0.0001	0.0006	0.0024	0.0081	0.0215	0.0472
	13	0.0000	0.0000	0.0000	0.0000	0.0000	0.0001	0.0005	0.0021	0.0068	0.0182
	14	0.0000	0.0000	0.0000	0.0000	0.0000	0.0000	0.0001	0.0004	0.0016	0.0052
	15	0.0000	0.0000	0.0000	0.0000	0.0000	0.0000	0.0000	0.0001	0.0003	0.0010
	16	0.0000	0.0000	0.0000	0.0000	0.0000	0.0000	0.0000	0.0000	0.0000	0.0001
	17	0.0000	0.0000	0.0000	0.0000	0.0000	0.0000	0.0000	0.0000	0.0000	0.0000
18	0	0.3972	0.1501	0.0536	0.0180	0.0056	0.0016	0.0004	0.0001	0.0000	0.0000
	1	0.3763	0.3002	0.1704	0.0811	0.0338	0.0126	0.0042	0.0012	0.0003	0.0001
	2	0.1683	0.2835	0.2556	0.1723	0.0958	0.0458	0.0190	0.0069	0.0022	0.0006
	3	0.0473	0.1680	0.2406	0.2297	0.1704	0.1046	0.0547	0.0246	0.0095	0.0031
	4	0.0093	0.0700	0.1592	0.2153	0.2130	0.1681	0.1104	0.0614	0.0291	0.0117
	5	0.0014	0.0218	0.0787	0.1507	0.1988	0.2017	0.1664	0.1146	0.0666	0.0327
	6	0.0002	0.0052	0.0301	0.0816	0.1436	0.1873	0.1941	0.1655	0.1181	0.0708
	7	0.0000	0.0010	0.0091	0.0350	0.0820	0.1376	0.1792	0.1892	0.1657	0.1214
	8	0.0000	0.0002	0.0022	0.0120	0.0376	0.0811	0.1327	0.1734	0.1864	0.1669
	9	0.0000	0.0000	0.0004	0.0033	0.0139	0.0386	0.0794	0.1284	0.1694	0.1855
	10	0.0000	0.0000	0.0001	0.0008	0.0042	0.0149	0.0385	0.0771	0.1248	0.1669
	11	0.0000	0.0000	0.0000	0.0001	0.0010	0.0046	0.0151	0.0374	0.0742	0.1214
	12	0.0000	0.0000	0.0000	0.0000	0.0002	0.0012	0.0047	0.0145	0.0354	0.0708
	13	0.0000	0.0000	0.0000	0.0000	0.0000	0.0002	0.0012	0.0045	0.0134	0.0327
	14	0.0000	0.0000	0.0000	0.0000	0.0000	0.0000	0.0002	0.0011	0.0039	0.0117
	15	0.0000	0.0000	0.0000	0.0000	0.0000	0.0000	0.0000	0.0002	0.0009	0.0031
	16	0.0000	0.0000	0.0000	0.0000	0.0000	0.0000	0.0000	0.0000	0.0001	0.0006
	17	0.0000	0.0000	0.0000	0.0000	0.0000	0.0000	0.0000	0.0000	0.0000	0.0001
	18	0.0000	0.0000	0.0000	0.0000	0.0000	0.0000	0.0000	0.0000	0.0000	0.0000
19	0	0.3774	0.1351	0.0456	0.0144	0.0042	0.0011	0.0003	0.0001	0.0000	0.0000
	1	0.3774	0.2852	0.1529	0.0685	0.0268	0.0093	0.0029	0.0008	0.0002	0.0000
	2	0.1787	0.2852	0.2428	0.1540	0.0803	0.0358	0.0138	0.0046	0.0013	0.0003
	3	0.0533	0.1796	0.2428	0.2182	0.1517	0.0869	0.0422	0.0175	0.0062	0.0018
	4	0.0112	0.0798	0.1714	0.2182	0.2023	0.1491	0.0909	0.0467	0.0203	0.0074
	5	0.0018	0.0266	0.0907	0.1636	0.2023	0.1916	0.1468	0.0933	0.0497	0.0222
	6	0.0002	0.0069	0.0374	0.0955	0.1574	0.1916	0.1844	0.1451	0.0949	0.0518
	7	0.0000	0.0014	0.0122	0.0443	0.0974	0.1525	0.1844	0.1797	0.1443	0.0961
	8	0.0000	0.0002	0.0032	0.0166	0.0487	0.0981	0.1489	0.1797	0.1771	0.1442
	9	0.0000	0.0000	0.0007	0.0051	0.0198	0.0514	0.0980	0.1464	0.1771	0.1762

Table T2 (continued)
BINOMIAL PROBABILITIES[a]

Instructions: See the Examples in Chapter 11

p

n	x	0.05	0.10	0.15	0.20	0.25	0.30	0.35	0.40	0.45	0.50
	10	0.0000	0.0000	0.0001	0.0013	0.0066	0.0220	0.0528	0.0976	0.1449	0.1762
	11	0.0000	0.0000	0.0000	0.0003	0.0018	0.0077	0.0233	0.0532	0.0970	0.1442
	12	0.0000	0.0000	0.0000	0.0000	0.0004	0.0022	0.0083	0.0237	0.0529	0.0961
	13	0.0000	0.0000	0.0000	0.0000	0.0001	0.0005	0.0024	0.0085	0.0233	0.0518
	14	0.0000	0.0000	0.0000	0.0000	0.0000	0.0001	0.0006	0.0024	0.0082	0.0222
	15	0.0000	0.0000	0.0000	0.0000	0.0000	0.0000	0.0001	0.0005	0.0022	0.0074
	16	0.0000	0.0000	0.0000	0.0000	0.0000	0.0000	0.0000	0.0001	0.0005	0.0018
	17	0.0000	0.0000	0.0000	0.0000	0.0000	0.0000	0.0000	0.0000	0.0001	0.0003
	18	0.0000	0.0000	0.0000	0.0000	0.0000	0.0000	0.0000	0.0000	0.0000	0.0000
	19	0.0000	0.0000	0.0000	0.0000	0.0000	0.0000	0.0000	0.0000	0.0000	0.0000
20	0	0.3585	0.1216	0.0388	0.0115	0.0032	0.0008	0.0002	0.0000	0.0000	0.0000
	1	0.3774	0.2702	0.1368	0.0576	0.0211	0.0068	0.0020	0.0005	0.0001	0.0000
	2	0.1887	0.2852	0.2293	0.1369	0.0669	0.0278	0.0100	0.0031	0.0008	0.0002
	3	0.0596	0.1901	0.2428	0.2054	0.1339	0.0716	0.0323	0.0123	0.0040	0.0011
	4	0.0133	0.0898	0.1821	0.2182	0.1897	0.1304	0.0738	0.0350	0.0139	0.0046
	5	0.0022	0.0319	0.1028	0.1746	0.2023	0.1789	0.1272	0.0746	0.0365	0.0148
	6	0.0003	0.0089	0.0454	0.1091	0.1686	0.1916	0.1712	0.1244	0.0746	0.0370
	7	0.0000	0.0020	0.0160	0.0545	0.1124	0.1643	0.1844	0.1659	0.1221	0.0739
	8	0.0000	0.0004	0.0046	0.0222	0.0609	0.1144	0.1614	0.1797	0.1623	0.1201
	9	0.0000	0.0001	0.0011	0.0074	0.0271	0.0654	0.1158	0.1597	0.1771	0.1602
	10	0.0000	0.0000	0.0002	0.0020	0.0099	0.0308	0.0686	0.1171	0.1593	0.1762
	11	0.0000	0.0000	0.0000	0.0005	0.0030	0.0120	0.0336	0.0710	0.1185	0.1602
	12	0.0000	0.0000	0.0000	0.0001	0.0008	0.0039	0.0136	0.0355	0.0727	0.1201
	13	0.0000	0.0000	0.0000	0.0000	0.0002	0.0010	0.0045	0.0146	0.0366	0.0739
	14	0.0000	0.0000	0.0000	0.0000	0.0000	0.0002	0.0012	0.0049	0.0150	0.0370
	15	0.0000	0.0000	0.0000	0.0000	0.0000	0.0000	0.0003	0.0013	0.0049	0.0148
	16	0.0000	0.0000	0.0000	0.0000	0.0000	0.0000	0.0000	0.0003	0.0013	0.0046
	17	0.0000	0.0000	0.0000	0.0000	0.0000	0.0000	0.0000	0.0000	0.0002	0.0011
	18	0.0000	0.0000	0.0000	0.0000	0.0000	0.0000	0.0000	0.0000	0.0000	0.0002
	19	0.0000	0.0000	0.0000	0.0000	0.0000	0.0000	0.0000	0.0000	0.0000	0.0000
	20	0.0000	0.0000	0.0000	0.0000	0.0000	0.0000	0.0000	0.0000	0.0000	0.0000

[a] Linear interpolations with respect to θ will in general be accurate at most to two decimal places.

Table T3
THE STANDARD NORMAL DISTRIBUTION

Instructions: See the Examples in Chapter 11

z	0.00	0.01	0.02	0.03	0.04	0.05	0.06	0.07	0.08	0.09
0.0	0.0000	0.0040	0.0080	0.0120	0.0160	0.0199	0.0239	0.0279	0.0319	0.0359
0.1	0.0398	0.0438	0.0478	0.0517	0.0557	0.0596	0.0636	0.0675	0.0714	0.0753
0.2	0.0793	0.0832	0.0871	0.0910	0.0948	0.0987	0.1026	0.1064	0.1103	0.1141
0.3	0.1179	0.1217	0.1255	0.1293	0.1331	0.1368	0.1406	0.1443	0.1480	0.1517
0.4	0.1554	0.1591	0.1628	0.1664	0.1700	0.1736	0.1772	0.1808	0.1844	0.1879
0.5	0.1915	0.1950	0.1985	0.2019	0.2054	0.2088	0.2123	0.2157	0.2190	0.2224
0.6	0.2257	0.2291	0.2324	0.2357	0.2389	0.2422	0.2454	0.2486	0.2517	0.2549
0.7	0.2580	0.2611	0.2642	0.2673	0.2704	0.2734	0.2764	0.2794	0.2823	0.2852
0.8	0.2881	0.2910	0.2939	0.2967	0.2995	0.3023	0.3051	0.3078	0.3106	0.3133
0.9	0.3159	0.3186	0.3212	0.3238	0.3264	0.3289	0.3315	0.3340	0.3365	0.3389
1.0	0.3413	0.3438	0.3461	0.3485	0.3508	0.3531	0.3554	0.3577	0.3599	0.3621
1.1	0.3643	0.3665	0.3686	0.3708	0.3729	0.3749	0.3770	0.3790	0.3810	0.3830
1.2	0.3849	0.3869	0.3888	0.3907	0.3925	0.3944	0.3962	0.3980	0.3997	0.4015
1.3	0.4032	0.4049	0.4066	0.4082	0.4099	0.4115	0.4131	0.4147	0.4162	0.4177
1.4	0.4192	0.4207	0.4222	0.4236	0.4251	0.4265	0.4279	0.4292	0.4306	0.4319
1.5	0.4332	0.4345	0.4357	0.4370	0.4382	0.4394	0.4406	0.4418	0.4429	0.4441
1.6	0.4452	0.4463	0.4474	0.4484	0.4495	0.4505	0.4515	0.4525	0.4535	0.4545
1.7	0.4554	0.4564	0.4573	0.4582	0.4591	0.4599	0.4608	0.4616	0.4625	0.4633
1.8	0.4641	0.4649	0.4656	0.4664	0.4671	0.4678	0.4686	0.4693	0.4699	0.4706
1.9	0.4713	0.4719	0.4726	0.4732	0.4738	0.4744	0.4750	0.4756	0.4761	0.4767
2.0	0.4772	0.4778	0.4783	0.4788	0.4793	0.4798	0.4803	0.4808	0.4812	0.4817
2.1	0.4821	0.4826	0.4830	0.4834	0.4838	0.4842	0.4846	0.4850	0.4854	0.4857
2.2	0.4861	0.4864	0.4868	0.4871	0.4875	0.4878	0.4881	0.4884	0.4887	0.4890
2.3	0.4893	0.4896	0.4898	0.4901	0.4904	0.4906	0.4909	0.4911	0.4913	0.4916
2.4	0.4918	0.4920	0.4922	0.4925	0.4927	0.4929	0.4931	0.4932	0.4934	0.4936
2.5	0.4938	0.4940	0.4941	0.4943	0.4945	0.4946	0.4948	0.4949	0.4951	0.4952
2.6	0.4953	0.4955	0.4956	0.4957	0.4959	0.4960	0.4961	0.4962	0.4963	0.4964
2.7	0.4965	0.4966	0.4967	0.4968	0.4969	0.4970	0.4971	0.4972	0.4973	0.4974
2.8	0.4974	0.4975	0.4976	0.4977	0.4977	0.4978	0.4979	0.4979	0.4980	0.4981
2.9	0.4981	0.4982	0.4982	0.4983	0.4984	0.4984	0.4985	0.4985	0.4986	0.4986
3.0	0.4987	0.4987	0.4987	0.4988	0.4988	0.4989	0.4989	0.4989	0.4990	0.4990

Table T4
UPPER α PROBABILITY POINTS OF STUDENT'S *t*
DISTRIBUTION (ENTRIES ARE $t_{a:\nu}$)

Instructions: (1)   Enter the row of the table corresponding to the number
                       of degrees of freedom (ν) for error.
                 (2)   Pick the value of t in that row, from the column that
                       corresponds to the predetermined α-level.

				α			
ν	0.25	0.10	0.05	0.025	0.01	0.005	0.0005
1	1.000	3.078	6.314	12.706	31.821	63.657	636.619
2	0.816	1.886	2.920	4.303	6.965	9.925	31.598
3	0.765	1.638	2.353	3.182	4.541	5.841	12.941
4	0.741	1.533	2.132	2.776	3.747	4.604	8.610
5	0.727	1.476	2.015	2.571	3.365	4.032	6.859
6	0.718	1.440	1.943	2.447	3.143	3.707	5.959
7	0.711	1.415	1.895	2.365	2.998	3.499	5.405
8	0.706	1.397	1.860	2.306	2.896	3.355	5.041
9	0.703	1.383	1.833	2.262	2.821	3.250	4.781
10	0.700	1.372	1.812	2.228	2.764	3.169	4.587
11	0.697	1.363	1.796	2.201	2.718	3.106	4.437
12	0.695	1.356	1.782	2.179	2.681	3.055	4.318
13	0.694	1.350	1.771	2.160	2.650	3.012	4.221
14	0.692	1.345	1.761	2.145	2.624	2.977	4.140
15	0.691	1.341	1.753	2.131	2.602	2.947	4.073
16	0.690	1.337	1.746	2.120	2.583	2.921	4.015
17	0.689	1.333	1.740	2.110	2.567	2.898	3.965
18	0.688	1.330	1.734	2.101	2.552	2.878	3.922
19	0.688	1.328	1.729	2.093	2.539	2.861	3.883
20	0.687	1.325	1.725	2.086	2.528	2.845	3.850
21	0.686	1.323	1.721	2.080	2.518	2.831	3.819
22	0.686	1.321	1.717	2.074	2.508	2.819	3.792
23	0.685	1.319	1.714	2.069	2.500	2.807	3.767
24	0.685	1.318	1.711	2.064	2.492	2.797	3.745
25	0.684	1.316	1.708	2.060	2.485	2.787	3.725
26	0.684	1.315	1.706	2.056	2.479	2.779	3.707
27	0.684	1.314	1.703	2.052	2.473	2.771	3.690
28	0.683	1.313	1.701	2.048	2.467	2.763	3.674
29	0.683	1.311	1.699	2.045	2.462	2.756	3.659
30	0.683	1.310	1.697	2.042	2.457	2.750	3.646
40	0.681	1.303	1.684	2.021	2.423	2.704	3.551
60	0.679	1.296	1.671	2.000	2.390	2.660	3.460
120	0.677	1.289	1.658	1.980	2.358	2.617	3.373
∞	0.674	1.282	1.645	1.960	2.326	2.576	3.291

This table is abridged from Table III of Fisher and Yates: *Statistical Tables for Biological, Agricultural, and Medical Research*, published by Longman Group Ltd., London (previously published by Oliver and Boyd, Edinburgh), and by permission of the authors and publishers.

Table T5
UPPER α PROBABILITY POINTS OF χ^2 DISTRIBUTION (ENTRIES ARE $\chi^2_{\alpha:\nu}$)

Instructions: (1) Enter the row of the table corresponding to the
number of degrees of freedom (ν) for χ^2.

(2) Pick the value of χ^2 in that row from the column
that corresponds to the predetermined α-level.

							α						
ν	0.995	0.990	0.975	0.950	0.900	0.750	0.500	0.250	0.100	0.050	0.025	0.010	0.005
1	0.0000393	0.000157	0.000982	0.00393	0.0158	0.102	0.455	1.32	2.71	3.84	5.02	6.63	7.88
2	0.0100	0.0201	0.0506	0.103	0.211	0.575	1.39	2.77	4.61	5.99	7.38	9.21	10.6
3	0.0717	0.115	0.216	0.352	0.584	1.21	2.37	4.11	6.25	7.81	9.35	11.3	12.8
4	0.207	0.297	0.484	0.711	1.06	1.92	3.36	5.39	7.78	9.49	11.1	13.3	14.9
5	0.412	0.554	0.831	1.15	1.61	2.67	4.35	6.63	9.24	11.1	12.8	15.1	16.7
6	0.676	.872	1.24	1.64	2.20	3.45	5.35	7.84	10.6	12.6	14.4	16.8	18.5
7	0.989	1.24	1.69	2.17	2.83	4.25	6.35	9.04	12.0	14.1	16.0	18.5	20.3
8	1.34	1.65	2.18	2.73	3.49	5.07	7.34	10.2	13.4	15.5	17.5	20.1	22.0
9	1.73	2.09	2.70	3.33	4.17	5.90	8.34	11.4	14.7	16.9	19.0	21.7	23.6
10	2.16	2.56	3.25	3.94	4.87	6.74	9.34	12.5	16.0	18.3	20.5	23.2	25.2
11	2.60	3.05	3.82	4.57	5.58	7.58	10.3	13.7	17.3	19.7	21.9	24.7	26.8
12	3.07	3.57	4.40	5.23	6.30	8.44	11.3	14.8	18.5	21.0	23.3	26.2	28.3
13	3.57	4.11	5.01	5.89	7.04	9.30	12.3	16.0	19.8	22.4	24.7	27.7	29.8
14	4.07	4.66	5.63	6.57	7.79	10.2	13.3	17.1	21.1	23.7	26.1	29.1	31.3
15	4.60	5.23	6.26	7.26	8.55	11.0	4.3	18.2	22.3	25.0	27.5	30.6	32.8
16	5.14	5.81	6.91	7.96	9.31	11.9	15.3	19.4	23.5	26.3	28.8	32.0	34.3
17	5.70	6.41	7.56	8.67	10.1	12.8	16.3	20.5	24.8	27.6	30.2	33.4	35.7
18	6.26	7.01	8.23	9.39	10.9	13.7	17.3	21.6	26.0	28.9	31.5	34.8	37.2
19	6.84	7.63	8.91	10.1	11.7	14.6	18.3	22.7	27.2	30.1	32.9	36.2	38.6
20	7.43	8.26	9.59	10.9	12.4	15.5	19.3	23.8	28.4	31.4	34.2	37.6	40.0
21	8.03	8.90	10.3	11.6	13.2	16.3	20.3	24.9	29.6	32.7	35.5	38.9	41.4
22	8.64	9.54	11.0	12.3	14.0	17.2	21.3	26.0	30.8	33.9	36.8	40.3	42.8
23	9.26	10.2	11.7	13.1	14.8	18.1	22.3	27.1	32.0	35.2	38.1	41.6	44.2
24	9.89	10.9	12.4	13.8	15.7	19.0	23.3	28.2	33.2	36.4	39.4	43.0	45.6
25	10.5	11.5	13.1	14.6	16.5	19.9	24.3	29.3	34.4	37.7	40.6	44.3	46.9
26	11.2	12.2	13.8	15.4	17.3	20.8	25.3	30.4	35.6	38.9	41.9	45.6	48.3
27	11.8	12.9	14.6	16.2	18.1	21.7	26.3	31.5	36.7	40.1	43.2	47.0	49.6
28	12.5	13.6	15.3	16.9	18.9	22.7	27.3	32.6	37.9	41.3	44.5	48.3	51.0
29	13.1	14.3	16.0	17.7	19.8	23.6	28.3	33.7	39.1	42.6	45.7	49.6	52.3
30	13.8	15.0	16.8	18.5	20.6	24.5	29.3	34.8	40.3	43.8	47.0	50.9	53.7

Table T6
UPPER α PROBABILITY POINTS OF F DISTRIBUTION
(ENTRIES ARE $F_{\alpha;\nu_1,\nu_2}$)

Instructions:

(1) Enter the section of the table corresponding to the predetermined α-level.

(2) Enter the row that corresponds to the denominator degrees-of-freedom (ν_2).

(3) Pick the value of F in that row from the column that corresponds to the numerator degrees-of-freedom (ν_1).

$\alpha = 0.10$

ν_2	1	2	3	4	5	6	7	8	9	10	12	15	20	24	30	40	60	120	∞
1	39.86	49.50	53.59	55.83	57.24	58.20	58.91	59.44	59.86	60.19	60.71	61.22	61.74	62.00	62.26	62.53	62.79	63.06	63.33
2	8.53	9.00	9.16	9.24	9.29	9.33	9.35	9.37	9.38	9.39	9.41	9.42	9.44	9.45	9.46	9.47	9.47	9.48	9.49
3	5.54	5.46	5.39	5.34	5.31	5.28	5.27	5.25	5.24	5.23	5.22	5.20	5.18	5.18	5.17	5.16	5.15	5.14	5.13
4	4.54	4.32	4.19	4.11	4.05	4.01	3.98	3.95	3.94	3.92	3.90	3.87	3.84	3.83	3.82	3.80	3.79	3.78	3.76
5	4.06	3.78	3.62	3.52	3.45	3.40	3.37	3.34	3.32	3.30	3.27	3.24	3.21	3.19	3.17	3.16	3.14	3.12	3.10
6	3.78	3.46	3.29	3.18	3.11	3.05	3.01	2.98	2.96	2.94	2.90	2.87	2.84	2.82	2.80	2.78	2.76	2.74	2.72
7	3.59	3.26	3.07	2.96	2.88	2.83	2.78	2.75	2.72	2.70	2.67	2.63	2.59	2.58	2.56	2.54	2.51	2.49	2.47
8	3.46	3.11	2.92	2.81	2.73	2.67	2.62	2.59	2.56	2.54	2.50	2.46	2.42	2.40	2.38	2.36	2.34	2.32	2.29
9	3.36	3.01	2.81	2.69	2.61	2.55	2.51	2.47	2.44	2.42	2.38	2.34	2.30	2.28	2.25	2.23	2.21	2.18	2.16
10	3.29	2.92	2.73	2.61	2.52	2.46	2.41	2.38	2.35	2.32	2.28	2.24	2.20	2.18	2.16	2.13	2.11	2.08	2.06
11	3.23	2.86	2.66	2.54	2.45	2.39	2.34	2.30	2.27	2.25	2.21	2.17	2.12	2.10	2.08	2.05	2.03	2.00	1.97
12	3.18	2.81	2.61	2.48	2.39	2.33	2.28	2.24	2.21	2.19	2.15	2.10	2.06	2.04	2.01	1.99	1.96	1.93	1.90
13	3.14	2.76	2.56	2.43	2.35	2.28	2.23	2.20	2.16	2.14	2.10	2.05	2.01	1.98	1.96	1.93	1.90	1.88	1.85
14	3.10	2.73	2.52	2.39	2.31	2.24	2.19	2.15	2.12	2.10	2.05	2.01	1.96	1.94	1.91	1.89	1.86	1.83	1.80
15	3.07	2.70	2.49	2.36	2.27	2.21	2.16	2.12	2.09	2.06	2.02	1.97	1.92	1.90	1.87	1.85	1.82	1.79	1.76
16	3.05	2.67	2.46	2.33	2.24	2.18	2.13	2.09	2.06	2.03	1.99	1.94	1.89	1.87	1.84	1.81	1.78	1.75	1.72

ν_1

Table T6 (continued)
UPPER α PROBABILITY POINTS OF F DISTRIBUTION
(ENTRIES ARE $F_{\alpha;\nu_1,\nu_2}$)

Instructions:

(1) Enter the section of the table corresponding to the predetermined α-level.
(2) Enter the row that corresponds to the denominator degrees-of-freedom (ν_2).
(3) Pick the value of F in that row from the column that corresponds to the numerator degrees-of-freedom (ν_1).

ν_2	ν_1 1	2	3	4	5	6	7	8	9	10	12	15	20	24	30	40	60	120	∞
17	3.03	2.64	2.44	2.31	2.22	2.15	2.10	2.06	2.03	2.00	1.96	1.91	1.86	1.84	1.81	1.78	1.75	1.72	1.69
18	3.01	2.62	2.42	2.29	2.20	2.13	2.08	2.04	2.00	1.98	1.93	1.89	1.84	1.81	1.78	1.75	1.72	1.69	1.66
19	2.99	2.61	2.40	2.27	2.18	2.11	2.06	2.02	1.98	1.96	1.91	1.86	1.81	1.79	1.76	1.73	1.70	1.67	1.63
20	2.97	2.59	2.38	2.25	2.16	2.09	2.04	2.00	1.96	1.94	1.89	1.84	1.79	1.77	1.74	1.71	1.68	1.64	1.61
21	2.96	2.57	2.36	2.23	2.14	2.08	2.02	1.98	1.95	1.92	1.87	1.83	1.78	1.75	1.72	1.69	1.66	1.62	1.59
22	2.95	2.56	2.35	2.22	2.13	2.06	2.01	1.97	1.93	1.90	1.86	1.81	1.76	1.73	1.70	1.67	1.64	1.60	1.57
23	2.94	2.55	2.34	2.21	2.11	2.05	1.99	1.95	1.92	1.89	1.84	1.80	1.74	1.72	1.69	1.66	1.62	1.59	1.55
24	2.93	2.54	2.33	2.19	2.10	2.04	1.98	1.94	1.91	1.88	1.83	1.78	1.73	1.70	1.67	1.64	1.61	1.57	1.53
25	2.92	2.53	2.32	2.18	2.09	2.02	1.97	1.93	1.89	1.87	1.82	1.77	1.72	1.69	1.66	1.63	1.59	1.56	1.52
26	2.91	2.52	2.31	2.17	2.08	2.01	1.96	1.92	1.88	1.86	1.81	1.76	1.71	1.68	1.65	1.61	1.58	1.54	1.50
27	2.90	2.51	2.30	2.17	2.07	2.00	1.95	1.91	1.87	1.85	1.80	1.75	1.70	1.67	1.64	1.60	1.57	1.53	1.49
28	2.89	2.50	2.29	2.16	2.06	2.00	1.94	1.90	1.87	1.84	1.79	1.74	1.69	1.66	1.63	1.59	1.56	1.52	1.48
29	2.89	2.50	2.28	2.15	2.06	1.99	1.93	1.89	1.86	1.83	1.78	1.73	1.68	1.65	1.62	1.58	1.55	1.51	1.47
30	2.88	2.49	2.28	2.14	2.05	1.98	1.93	1.88	1.85	1.82	1.77	1.72	1.67	1.64	1.61	1.57	1.54	1.50	1.46
40	2.84	2.44	2.23	2.09	2.00	1.93	1.87	1.83	1.79	1.76	1.71	1.66	1.61	1.57	1.54	1.51	1.47	1.42	1.38
60	2.79	2.39	2.18	2.04	1.95	1.87	1.82	1.77	1.74	1.71	1.66	1.60	1.54	1.51	1.48	1.44	1.40	1.35	1.29
120	2.75	2.35	2.13	1.99	1.90	1.82	1.77	1.72	1.68	1.65	1.60	1.55	1.48	1.45	1.41	1.37	1.32	1.26	1.19

α = 0.05

	1.00	1.17	1.24	1.30	1.34	1.38	1.42	1.49	1.55	1.60	1.63	1.67	1.72	1.77	1.85	1.94	2.08	2.30	2.71
1	254.3	253.3	252.2	251.1	250.1	249.1	248.0	245.9	243.9	241.9	240.5	238.9	236.8	234.0	230.2	224.6	215.7	199.5	161.4
2	19.50	19.49	19.48	19.47	19.46	19.45	19.45	19.43	19.41	19.40	19.38	19.37	19.35	19.33	19.30	19.25	19.16	19.00	18.51
3	8.53	8.55	8.57	8.59	8.62	8.64	8.66	8.70	8.74	8.79	8.81	8.85	8.89	8.94	9.01	9.12	9.28	9.55	10.13
4	5.63	5.66	5.69	5.72	5.75	5.77	5.80	5.86	5.91	5.95	6.00	6.04	6.09	6.16	6.26	6.39	6.59	6.94	7.71
5	4.36	4.40	4.43	4.46	4.50	4.53	4.56	4.62	4.68	4.74	4.77	4.82	4.88	4.95	5.05	5.19	5.41	5.79	6.61
6	3.67	3.70	3.74	3.77	3.81	3.84	3.87	3.94	4.00	4.06	4.10	4.15	4.21	4.28	4.39	4.53	4.76	5.14	5.99
7	3.23	3.27	3.30	3.34	3.38	3.41	3.44	3.51	3.57	3.64	3.68	3.73	3.79	3.87	3.97	4.12	4.35	4.74	5.59
8	2.93	2.97	3.01	3.04	3.08	3.12	3.15	3.22	3.28	3.35	3.39	3.44	3.50	3.58	3.69	3.84	4.07	4.46	5.32
9	2.71	2.75	2.79	2.83	2.86	2.90	2.94	3.01.	3.07	3.14	3.18	3.23	3.29	3.37	3.48	3.63	3.86	4.26	5.12
10	2.54	2.58	2.62	2.66	2.70	2.74	2.77	2.85	2.91	2.98	3.02	3.07	3.14	3.22	3.33	3.48	3.71	4.10	4.96
11	2.40	2.45	2.49	2.53	2.57	2.61	2.65	2.72	2.79	2.85	2.90	2.95	3.01	3.09	3.20	3.36	3.59	3.98	4.84
12	2.30	2.34	2.38	2.43	2.47	2.51	2.54	2.62	2.69	2.75	2.80	2.85	2.91	3.00	3.11	3.26	3.49	3.89	4.75
13	2.21	2.25	2.30	2.34	2.38	2.42	2.46	2.53	2.60	2.67	2.71	2.77	2.83	2.92	3.03	3.18	3.41	3.81	4.67
14	2.13	2.18	2.22	2.27	2.31	2.35	2.39	2.46	2.53	2.60	2.65	2.70	2.76	2.85	2.96	3.11	3.34	3.74	4.60
15	2.07	2.11	2.16	2.20	2.25	2.29	2.33	2.40	2.48	2.54	2.59	2.64	2.71	2.79	2.90	3.06	3.29	3.68	4.54
16	2.01	2.06	2.11	2.15	2.19	2.24	2.28	2.35	2.42	2.49	2.54	2.59	2.66	2.74	2.85	3.01	3.24	3.63	4.49
17	1.96	2.01	2.06	2.10	2.15	2.19	2.23	2.31	2.38	2.45	2.49	2.55	2.61	2.70	2.81	2.96	3.20	3.59	4.45
18	1.92	1.97	2.02	2.06	2.11	2.15	2.19	2.27	2.34	2.41	2.46	2.51	2.58	2.66	2.77	2.93	3.16	3.55	4.41
19	1.88	1.93	1.98	2.03	2.07	2.11	2.16	2.23	2.31	2.38	2.42	2.48	2.54	2.63	2.74	2.90	3.13	3.52	4.38
20	1.84	1.90	1.95	1.99	2.04	2.08	2.12	2.20	2.28	2.35	2.39	2.45	2.51	2.60	2.71	2.87	3.10	3.49	4.35
21	1.81	1.87	1.92	1.96	2.01	2.05	2.10	2.18	2.25	2.32	2.37	2.42	2.49	2.57	2.68	2.84	3.07	3.47	4.32
22	1.78	1.84	1.89	1.94	1.98	2.03	2.07	2.15	2.23	2.30	2.34	2.40	2.46	2.55	2.66	2.82	3.05	3.44	4.30
23	1.76	1.81	1.86	1.91	1.96	2.01	2.05	2.13	2.20	2.27	2.32	2.37	2.44	2.53	2.64	2.80	3.03	3.42	4.28
24	1.73	1.79	1.84	1.89	1.94	1.98	2.03	2.11	2.18	2.25	2.30	2.36	2.42	2.51	2.62	2.78	3.01	3.40	4.26
25	1.71	1.77	1.82	1.87	1.92	1.96	2.01	2.09	2.16	2.24	2.28	2.34	2.40	2.49	2.60	2.76	2.99	3.39	4.24
26	1.69	1.75	1.80	1.85	1.90	1.95	1.99	2.07	2.15	2.22	2.27	2.32	2.39	2.47	2.59	2.74	2.98	3.37	4.23
27	1.67	1.73	1.79	1.84	1.88	1.93	1.97	2.06	2.13	2.20	2.25	2.31	2.37	2.46	2.57	2.73	2.96	3.35	4.21
28	1.65	1.71	1.77	1.82	1.87	1.91	1.96	2.04	2.12	2.19	2.24	2.29	2.36	2.45	2.56	2.71	2.95	3.34	4.20

Table T6 (continued)
UPPER α PROBABILITY POINTS OF F DISTRIBUTION
(ENTRIES ARE $F_{\alpha:\nu_1,\nu_2}$)

Instructions:
(1) Enter the section of the table corresponding to the predetermined α-level.
(2) Enter the row that corresponds to the denominator degrees-of-freedom (ν_2).
(3) Pick the value of F in that row from the column that corresponds to the numerator degrees-of-freedom (ν_1).

ν_2	ν_1 1	2	3	4	5	6	7	8	9	10	12	15	20	24	30	40	60	120	∞
29	4.18	3.33	2.93	2.70	2.55	2.43	2.35	2.28	2.22	2.18	2.10	2.03	1.94	1.90	1.85	1.81	1.75	1.70	1.64
30	4.17	3.32	2.92	2.69	2.53	2.42	2.33	2.27	2.21	2.16	2.09	2.01	1.93	1.89	1.84	1.79	1.74	1.68	1.62
40	4.08	3.23	2.84	2.61	2.45	2.34	2.25	2.18	2.12	2.08	2.00	1.92	1.84	1.79	1.74	1.69	1.64	1.58	1.51
60	4.00	3.15	2.76	2.53	2.37	2.25	2.17	2.10	2.04	1.99	1.92	1.84	1.75	1.70	1.65	1.59	1.53	1.47	1.39
120	3.92	3.07	2.68	2.45	2.29	2.17	2.09	2.02	1.96	1.91	1.83	1.75	1.66	1.61	1.55	1.50	1.43	1.35	1.25
∞	3.84	3.00	2.60	2.37	2.21	2.10	2.01	1.94	1.88	1.83	1.75	1.67	1.57	1.52	1.46	1.39	1.32	1.22	1.00

$\alpha = 0.025$

ν_2	1	2	3	4	5	6	7	8	9	10	12	15	20	24	30	40	60	120	∞
1	647.8	799.5	864.2	899.6	921.8	937.1	948.2	956.7	963.3	968.6	976.7	984.9	993.1	997.2	1001	1006	1010	1014	1018
2	38.51	39.00	39.17	39.25	39.30	39.33	39.36	39.37	39.39	39.40	39.41	39.43	39.45	39.46	39.46	39.47	39.48	39.49	39.50
3	17.44	16.04	15.44	15.10	14.88	14.73	14.62	14.54	14.47	14.42	14.34	14.25	14.17	14.12	14.08	14.04	13.99	13.95	13.90
4	12.22	10.65	9.98	9.60	9.36	9.20	9.07	8.98	8.90	8.84	8.75	8.66	8.56	8.51	8.46	8.41	8.36	8.31	8.26
5	10.01	8.43	7.76	7.39	7.15	6.98	6.85	6.76	6.68	6.62	6.52	6.43	6.33	6.28	6.23	6.18	6.12	6.07	6.02
6	8.81	7.26	6.60	6.23	5.99	5.82	5.70	5.60	5.52	5.46	5.37	5.27	5.17	5.12	5.07	5.01	4.96	4.90	4.85
7	8.07	6.54	5.89	5.52	5.29	5.12	4.99	4.90	4.82	4.76	4.67	4.57	4.47	4.42	4.36	4.31	4.25	4.20	4.14
8	7.57	6.06	5.42	5.05	4.82	4.65	4.53	4.43	4.36	4.30	4.20	4.10	4.00	3.95	3.89	3.84	3.78	3.73	3.67
9	7.21	5.71	5.08	4.72	4.48	4.32	4.20	4.10	4.03	3.96	3.87	3.77	3.67	3.61	3.56	3.51	3.45	3.39	3.33

10	6.94	5.46	4.83	4.47	4.24	4.07	3.95	3.85	3.78	3.72	3.62	3.52	3.42	3.37	3.31	3.26	3.20	3.14	3.08
11	6.72	5.26	4.63	4.28	4.04	3.88	3.76	3.66	3.59	3.53	3.43	3.33	3.23	3.17	3.12	3.06	3.00	2.94	2.88
12	6.55	5.10	4.47	4.12	3.89	3.73	3.61	3.51	3.44	3.37	3.28	3.18	3.07	3.02	2.96	2.91	2.85	2.79	2.72
13	6.41	4.97	4.35	4.00	3.77	3.60	3.48	3.39	3.31	3.25	3.15	3.05	2.95	2.89	2.84	2.78	2.72	2.66	2.60
14	6.30	4.86	4.24	3.89	3.66	3.50	3.38	3.29	3.21	3.15	3.05	2.95	2.84	2.79	2.73	2.67	2.61	2.55	2.49
15	6.20	4.77	4.15	3.80	3.58	3.41	3.29	3.20	3.12	3.06	2.96	2.86	2.76	2.70	2.64	2.59	2.52	2.46	2.40
16	6.12	4.69	4.08	3.73	3.50	3.34	3.22	3.12	3.05	2.99	2.89	2.79	2.68	2.63	2.57	2.51	2.45	2.38	2.32
17	6.04	4.62	4.01	3.66	3.44	3.28	3.16	3.06	2.98	2.92	2.82	2.72	2.62	2.56	2.50	2.44	2.38	2.32	2.25
18	5.98	4.56	3.95	3.61	3.38	3.22	3.10	3.01	2.93	2.87	2.77	2.67	2.56	2.50	2.44	2.38	2.32	2.26	2.19
19	5.92	4.51	3.90	3.56	3.33	3.17	3.05	2.96	2.88	2.82	2.72	2.62	2.51	2.45	2.39	2.33	2.27	2.20	2.13
20	5.87	4.46	3.86	3.51	3.29	3.13	3.01	2.91	2.84	2.77	2.68	2.57	2.46	2.41	2.35	2.29	2.22	2.16	2.09
21	5.83	4.42	3.82	3.48	3.25	3.09	2.97	2.87	2.80	2.73	2.64	2.53	2.42	2.37	2.31	2.25	2.18	2.11	2.04
22	5.79	4.38	3.78	3.44	3.22	3.05	2.93	2.84	2.76	2.70	2.60	2.50	2.39	2.33	2.27	2.21	2.14	2.08	2.00
23	5.75	4.35	3.75	3.41	3.18	3.02	2.90	2.81	2.73	2.67	2.57	2.47	2.36	2.30	2.24	2.18	2.11	2.04	1.97
24	5.72	4.32	3.72	3.38	3.15	2.99	2.87	2.78	2.70	2.64	2.54	2.44	2.33	2.27	2.21	2.15	2.08	2.01	1.94
25	5.69	4.29	3.69	3.35	3.13	2.97	2.85	2.75	2.68	2.61	2.51	2.41	2.30	2.24	2.18	2.12	2.05	1.98	1.91
26	5.66	4.27	3.67	3.33	3.10	2.94	2.82	2.73	2.65	2.59	2.49	2.39	2.28	2.22	2.16	2.09	2.03	1.95	1.88
27	5.63	4.24	3.65	3.31	3.08	2.92	2.80	2.71	2.63	2.57	2.47	2.36	2.25	2.19	2.13	2.07	2.00	1.93	1.85
28	5.61	4.22	3.63	3.29	3.06	2.90	2.78	2.69	2.61	2.55	2.45	2.34	2.23	2.17	2.11	2.05	1.98	1.91	1.83
29	5.59	4.20	3.61	3.27	3.04	2.88	2.76	2.67	2.59	2.53	2.43	2.32	2.21	2.15	2.09	2.03	1.96	1.89	1.81
30	5.57	4.18	3.59	3.25	3.03	2.87	2.75	2.65	2.57	2.51	2.41	2.31	2.20	2.14	2.07	2.01	1.94	1.87	1.79
40	5.42	4.05	3.46	3.13	2.90	2.74	2.62	2.53	2.45	2.39	2.29	2.18	2.07	2.01	1.94	1.88	1.80	1.72	1.64
60	5.29	3.93	3.34	3.01	2.79	2.63	2.51	2.41	2.33	2.27	2.17	2.06	1.94	1.88	1.82	1.74	1.67	1.58	1.48
120	5.15	3.80	3.23	2.89	2.67	2.52	2.39	2.30	2.22	2.16	2.05	1.94	1.82	1.76	1.69	1.61	1.53	1.43	1.31
∞	5.02	3.69	3.12	2.79	2.57	2.41	2.29	2.19	2.11	2.05	1.94	1.83	1.71	1.64	1.57	1.48	1.39	1.27	1.00

$\alpha = 0.01$

1	4052	4999.5	5403	5625	5764	5859	5928	5982	6022	6056	6106	6157	6209	6235	6261	6287	6313	6339	6366
2	98.50	99.00	99.17	99.25	99.30	99.33	99.36	99.37	99.39	99.40	99.42	99.43	99.45	99.46	99.47	99.47	99.48	99.49	99.50
3	34.12	30.82	29.46	28.71	28.24	27.91	27.67	27.49	27.35	27.23	27.05	26.87	26.69	26.60	26.50	26.41	26.32	26.22	26.13
4	21.20	18.00	16.69	15.98	15.52	15.21	14.98	14.80	14.65	14.55	14.37	14.20	14.02	13.93	13.84	13.75	13.65	13.56	13.46

Table T6 (continued)
UPPER α PROBABILITY POINTS OF F DISTRIBUTION
(ENTRIES ARE $F_{\alpha; \nu_1, \nu_2}$)

Instructions:

(1) Enter the section of the table corresponding to the predetermined α-level.
(2) Enter the row that corresponds to the denominator degrees-of-freedom (ν_2).
(3) Pick the value of F in that row from the column that corresponds to the numerator degrees-of-freedom (ν_1).

ν_1

ν_2	1	2	3	4	5	6	7	8	9	10	12	15	20	24	30	40	60	120	∞
5	16.26	13.27	12.06	11.39	10.97	10.67	10.46	10.29	10.16	10.05	9.89	9.72	9.55	9.47	9.38	9.29	9.20	9.11	9.02
6	13.75	10.92	9.78	9.15	8.75	8.47	8.26	8.10	7.98	7.87	7.72	7.56	7.40	7.31	7.23	7.14	7.06	6.97	6.88
7	12.25	9.55	8.45	7.85	7.46	7.19	6.99	6.84	6.72	6.62	6.47	6.31	6.16	6.07	5.99	5.91	5.82	5.74	5.65
8	11.26	8.65	7.59	7.01	6.63	6.37	6.18	6.03	5.91	5.81	5.67	5.52	5.36	5.28	5.20	5.12	5.03	4.95	4.86
9	10.56	8.02	6.99	6.42	6.06	5.80	5.61	5.47	5.35	5.26	5.11	4.96	4.81	4.73	4.65	4.57	4.48	4.40	4.31
10	10.04	7.56	6.55	5.99	5.64	5.39	5.20	5.06	4.94	4.85	4.71	4.56	4.41	4.33	4.25	4.17	4.08	4.00	3.91
11	9.65	7.21	6.22	5.67	5.32	5.07	4.89	4.74	4.63	4.54	4.40	4.25	4.10	4.02	3.94	3.86	3.78	3.69	3.60
12	9.33	6.93	5.95	5.41	5.06	4.82	4.64	4.50	4.39	4.30	4.16	4.01	3.86	3.78	3.70	3.62	3.54	3.45	3.36
13	9.07	6.70	5.74	5.21	4.86	4.62	4.44	4.30	4.19	4.10	3.96	3.82	3.66	3.59	3.51	3.43	3.34	3.25	3.17
14	8.86	6.51	5.56	5.04	4.69	4.46	4.28	4.14	4.03	3.94	3.80	3.66	3.51	3.43	3.35	3.27	3.18	3.09	3.00
15	8.68	6.36	5.42	4.89	4.56	4.32	4.14	4.00	3.89	3.80	3.67	3.52	3.37	3.29	3.21	3.13	3.05	2.96	2.87
16	8.53	6.23	5.29	4.77	4.44	4.20	4.03	3.89	3.78	3.69	3.55	3.41	3.26	3.18	3.10	3.02	2.93	2.84	2.75
17	8.40	6.11	5.18	4.67	4.34	4.10	3.93	3.79	3.68	3.59	3.46	3.31	3.16	3.08	3.00	2.92	2.83	2.75	2.65
18	8.29	6.01	5.09	4.58	4.25	4.01	3.84	3.71	3.60	3.51	3.37	3.23	3.08	3.00	2.92	2.84	2.75	2.66	2.57
19	8.18	5.93	5.01	4.50	4.17	3.94	3.77	3.63	3.52	3.43	3.30	3.15	3.00	2.92	2.84	2.76	2.67	2.58	2.49
20	8.10	5.85	4.94	4.43	4.10	3.87	3.70	3.56	3.46	3.37	3.23	3.09	2.94	2.86	2.78	2.69	2.61	2.52	2.42
21	8.02	5.78	4.87	4.37	4.04	3.81	3.64	3.51	3.40	3.31	3.17	3.03	2.88	2.80	2.72	2.64	2.55	2.46	2.36

22	7.95	5.72	4.82	4.31	3.99	3.76	3.59	3.45	3.35	3.26	3.12	2.98	2.83	2.75	2.67	2.58	2.50	2.40	2.31
23	7.88	5.66	4.76	4.26	3.94	3.71	3.54	3.41	3.30	3.21	3.07	2.93	2.78	2.70	2.62	2.54	2.45	2.35	2.26
24	7.82	5.61	4.72	4.22	3.90	3.67	3.50	3.36	3.26	3.17	3.03	2.89	2.74	2.66	2.58	2.49	2.40	2.31	2.21
25	7.77	5.57	4.68	4.18	3.85	3.63	3.46	3.32	3.22	3.13	2.99	2.85	2.70	2.62	2.54	2.45	2.36	2.27	2.17
26	7.72	5.53	4.64	4.14	3.82	3.59	3.42	3.29	3.18	3.09	2.96	2.81	2.66	2.58	2.50	2.42	2.33	2.23	2.13
27	7.68	5.49	4.60	4.11	3.78	3.56	3.39	3.26	3.15	3.06	2.93	2.78	2.63	2.55	2.47	2.38	2.29	2.20	2.10
28	7.64	5.45	4.57	4.07	3.75	3.53	3.36	3.23	3.12	3.03	2.90	2.75	2.60	2.52	2.44	2.35	2.26	2.17	2.06
29	7.60	5.42	4.54	4.04	3.73	3.50	3.33	3.20	3.09	3.00	2.87	2.73	2.57	2.49	2.41	2.33	2.23	2.14	2.03
30	7.56	5.39	4.51	4.02	3.70	3.47	3.30	3.17	3.07	2.98	2.84	2.70	2.55	2.47	2.39	2.30	2.21	2.11	2.01
40	7.31	5.18	4.31	3.83	3.51	3.29	3.12	2.99	2.89	2.80	2.66	2.52	2.37	2.29	2.20	2.11	2.02	1.92	1.80
60	7.08	4.98	4.13	3.65	3.34	3.12	2.95	2.82	2.72	2.63	2.50	2.35	2.20	2.12	2.03	1.94	1.84	1.73	1.60
120	6.85	4.79	3.95	3.48	3.17	2.96	2.79	2.66	2.56	2.47	2.34	2.19	2.03	1.95	1.86	1.76	1.66	1.53	1.38
∞	6.63	4.61	3.78	3.32	3.02	2.80	2.64	2.51	2.41	2.32	2.18	2.04	1.88	1.79	1.70	1.59	1.47	1.32	1.00

$$\alpha = 0.005$$

1	16211	20000	21615	22500	23056	23437	23715	23925	24091	24224	24426	24630	24836	24940	25044	25148	25253	25359	25465
2	198.5	199.0	199.2	199.2	199.3	199.3	199.4	199.4	199.4	199.4	199.4	199.4	199.4	199.5	199.5	199.5	199.5	199.5	199.5
3	55.55	49.80	47.47	46.19	45.39	44.84	44.43	44.13	43.88	43.69	43.39	43.08	42.78	42.62	42.47	42.31	42.15	41.99	41.83
4	31.33	26.28	24.26	23.15	22.46	21.97	21.62	21.35	21.14	20.97	20.70	20.44	20.17	20.03	19.89	19.75	19.61	19.47	19.32
5	22.78	18.31	16.53	15.56	14.94	14.51	14.20	13.96	13.77	13.62	13.38	13.15	12.90	12.78	12.66	12.53	12.40	12.27	12.14
6	18.63	14.54	12.92	12.03	11.46	11.07	10.79	10.57	10.39	10.25	10.03	9.81	9.59	9.47	9.36	9.24	9.12	9.00	8.88
7	16.24	12.40	10.88	10.05	9.52	9.16	8.89	8.68	8.51	8.38	8.18	7.97	7.75	7.65	7.53	7.42	7.31	7.19	7.08
8	14.69	11.04	9.60	8.81	8.30	7.95	7.69	7.50	7.34	7.21	7.01	6.81	6.61	6.50	6.40	6.29	6.18	6.06	5.95
9	13.61	10.11	8.72	7.96	7.47	7.13	6.88	6.69	6.54	6.42	6.23	6.03	5.83	5.73	5.62	5.52	5.41	5.30	5.19
10	12.83	9.43	8.08	7.34	6.87	6.54	6.30	6.12	5.97	5.85	5.66	5.47	5.27	5.17	5.07	4.97	4.86	4.75	4.64
11	12.23	8.91	7.60	6.88	6.42	6.10	5.86	5.68	5.54	5.42	5.24	5.05	4.86	4.76	4.65	4.55	4.44	4.34	4.23
12	11.75	8.51	7.23	6.52	6.07	5.76	5.52	5.35	5.20	5.09	4.91	4.72	4.53	4.43	4.33	4.23	4.12	4.01	3.90
13	11.37	8.19	6.93	6.23	5.79	5.48	5.25	5.08	4.94	4.82	4.64	4.46	4.27	4.17	4.07	3.97	3.87	3.76	3.65
14	11.06	7.92	6.68	6.00	5.56	5.26	5.03	4.86	4.72	4.60	4.43	4.25	4.06	3.96	3.86	3.76	3.66	3.55	3.44
15	10.80	7.70	6.48	5.80	5.37	5.07	4.85	4.67	4.54	4.42	4.25	4.07	3.88	3.79	3.69	3.58	3.48	3.37	3.26
16	10.58	7.51	6.30	5.64	5.21	4.91	4.69	4.52	4.33	4.27	4.10	3.92	3.73	3.64	3.54	3.44	3.33	3.22	3.11

Table T6 (continued)
UPPER α PROBABILITY POINTS OF F DISTRIBUTION
(ENTRIES ARE $F_{\alpha;\nu_1,\nu_2}$)

Instructions:

(1) Enter the section of the table corresponding to the predetermined α-level.
(2) Enter the row that corresponds to the denominator degrees-of-freedom (ν_2).
(3) Pick the value of F in that row from the column that corresponds to the numerator degrees-of-freedom (ν_1).

ν_2	ν_1 1	2	3	4	5	6	7	8	9	10	12	15	20	24	30	40	60	120	∞
17	10.38	7.35	6.16	5.50	5.07	4.78	4.56	4.39	4.25	4.14	3.97	3.79	3.61	3.51	3.41	3.31	3.21	3.10	2.98
18	10.22	7.21	6.03	5.37	4.96	4.66	4.44	4.28	4.14	4.03	3.86	3.68	3.50	3.40	3.30	3.20	3.10	2.99	2.87
19	10.07	7.09	5.92	5.27	4.85	4.56	4.34	4.18	4.04	3.93	3.76	3.59	3.40	3.31	3.21	3.11	3.00	2.89	2.78
20	9.94	6.99	5.82	5.17	4.76	4.47	4.26	4.09	3.96	3.85	3.68	3.50	3.32	3.22	3.12	3.02	2.92	2.81	2.69
21	9.83	6.89	5.73	5.09	4.68	4.39	4.18	4.01	3.88	3.77	3.60	3.43	3.24	3.15	3.05	2.95	2.84	2.73	2.61
22	9.73	6.81	5.65	5.02	4.61	4.32	4.11	3.94	3.81	3.70	3.54	3.36	3.18	3.08	2.98	2.88	2.77	2.66	2.55
23	9.63	6.73	5.58	4.95	4.54	4.26	4.05	3.88	3.75	3.64	3.47	3.30	3.12	3.02	2.92	2.82	2.71	2.60	2.48
24	9.55	6.66	5.52	4.89	4.49	4.20	3.99	3.83	3.69	3.59	3.42	3.25	3.06	2.97	2.87	2.77	2.66	2.55	2.43
25	9.48	6.60	5.46	4.84	4.43	4.15	3.94	3.78	3.64	3.54	3.37	3.20	3.01	2.92	2.82	2.72	2.61	2.50	2.38
26	9.41	6.54	5.41	4.79	4.38	4.10	3.89	3.73	3.60	3.49	3.33	3.15	2.97	2.87	2.77	2.67	2.56	2.45	2.33
27	9.34	6.49	5.36	4.74	4.34	4.06	3.85	3.69	3.56	3.45	3.28	3.11	2.93	2.83	2.73	2.63	2.52	2.41	2.25
28	9.28	6.44	5.32	4.70	4.30	4.02	3.81	3.65	3.52	3.41	3.25	3.07	2.89	2.79	2.69	2.59	2.48	2.37	2.29
29	9.23	6.40	5.28	4.66	4.26	3.98	3.77	3.61	3.48	3.38	3.21	3.04	2.86	2.76	2.66	2.56	2.45	2.33	2.24
30	9.18	6.35	5.24	4.62	4.23	3.95	3.74	3.58	3.45	3.34	3.18	3.01	2.82	2.73	2.63	2.52	2.42	2.30	2.18
40	8.83	6.07	4.98	4.37	3.99	3.71	3.51	3.35	3.22	3.12	2.95	2.78	2.60	2.50	2.40	2.30	2.18	2.06	1.93
60	8.49	5.79	4.73	4.14	3.76	3.49	3.29	3.13	3.01	2.90	2.74	2.57	2.39	2.29	2.19	2.08	1.96	1.83	1.69

	1	2	3	4	5	6	7	8	9	10	12	15	20	24	30	40	60	120	∞
120	8.18	5.54	4.50	3.92	3.55	3.28	3.09	2.93	2.81	2.71	2.54	2.37	2.19	2.09	1.98	1.87	1.75	1.61	1.43
∞	7.88	5.30	4.28	3.72	3.35	3.09	2.90	2.74	2.62	2.52	2.36	2.19	2.00	1.90	1.79	1.67	1.53	1.36	1.00

$\alpha = 0.001$

	1	2	3	4	5	6	7	8	9	10	12	15	20	24	30	40	60	120	∞
1	4053[a]	5000[a]	5404[a]	5625[a]	5764[a]	5859[a]	5929[a]	5981[a]	6023[a]	6056[a]	6107[a]	6158[a]	6209[a]	6235[a]	6261[a]	6287[a]	6313[a]	6340[a]	6366[a]
2	998.5	999.0	999.2	999.2	999.3	999.3	999.4	999.4	999.4	999.4	999.4	999.4	999.4	999.5	999.5	999.5	999.5	999.5	999.5
3	167.0	148.5	141.1	137.1	134.6	132.8	131.6	130.6	129.9	129.2	128.3	127.4	126.4	125.9	125.4	125.0	124.5	124.0	123.5
4	74.14	61.25	56.18	53.44	51.71	50.53	49.66	49.00	48.47	48.05	47.41	46.76	46.10	45.77	45.43	45.09	44.75	44.40	44.05
5	47.18	37.12	33.20	31.09	29.75	28.84	28.16	27.64	27.24	26.92	26.42	25.91	25.39	25.14	24.87	24.60	24.33	24.06	23.79
6	35.51	27.00	23.70	21.92	20.81	20.03	19.46	19.03	18.69	18.41	17.99	17.56	17.12	16.89	16.67	16.44	16.21	15.99	15.75
7	29.25	21.69	18.77	17.19	16.21	15.52	15.02	14.63	14.33	14.08	13.71	13.32	12.93	12.73	12.53	12.33	12.12	11.91	11.70
8	25.42	18.49	15.83	14.39	13.49	12.86	12.40	12.04	11.77	11.54	11.19	10.84	10.48	10.30	10.11	9.92	9.73	9.53	9.33
9	22.86	16.39	13.90	12.56	11.71	11.13	10.70	10.37	10.11	9.89	9.57	9.24	8.90	8.72	8.55	8.37	8.19	8.00	7.81
10	21.04	14.91	12.55	11.28	10.48	9.92	9.52	9.20	8.96	8.75	8.45	8.13	7.80	7.64	7.47	7.30	7.12	6.94	6.76
11	19.69	13.81	11.56	10.35	9.58	9.05	8.66	8.35	8.12	7.92	7.63	7.32	7.01	6.85	6.68	6.52	6.35	6.17	6.00
12	18.64	12.97	10.80	9.63	8.89	8.38	8.00	7.71	7.48	7.29	7.00	6.71	6.40	6.25	6.09	5.93	5.76	5.59	5.42
13	17.81	12.31	10.21	9.07	8.35	7.86	7.49	7.21	6.98	6.80	6.52	6.23	5.93	5.78	5.63	5.47	5.30	5.14	4.97
14	17.14	11.78	9.73	8.62	7.92	7.43	7.08	6.80	6.58	6.40	6.13	5.85	5.56	5.41	5.25	5.10	4.94	4.77	4.60
15	16.59	11.34	9.34	8.25	7.57	7.09	6.74	6.47	6.26	6.08	5.81	5.54	5.25	5.10	4.95	4.80	4.64	4.47	4.31
16	16.12	10.97	9.00	7.94	7.27	6.81	6.46	6.19	5.98	5.81	5.55	5.27	4.99	4.85	4.70	4.54	4.39	4.23	4.06
17	15.72	10.66	8.73	7.68	7.02	6.56	6.22	5.96	5.75	5.58	5.32	5.05	4.78	4.63	4.48	4.33	4.18	4.02	3.85
18	15.38	10.39	8.49	7.46	6.81	6.35	6.02	5.76	5.56	5.39	5.13	4.87	4.59	4.45	4.30	4.15	4.00	3.84	3.67
19	15.08	10.16	8.28	7.26	6.62	6.18	5.85	5.59	5.39	5.22	4.97	4.70	4.43	4.29	4.14	3.99	3.84	3.68	3.51
20	14.82	9.95	8.10	7.10	6.46	6.02	5.69	5.44	5.24	5.08	4.82	4.56	4.29	4.15	4.00	3.86	3.70	3.54	3.38
21	14.59	9.77	7.94	6.95	6.32	5.88	5.56	5.31	5.11	4.95	4.70	4.44	4.17	4.03	3.88	3.74	3.58	3.42	3.26
22	14.38	9.61	7.80	6.81	6.19	5.76	5.44	5.19	4.99	4.83	4.58	4.33	4.06	3.92	3.78	3.63	3.48	3.32	3.15
23	14.19	9.47	7.67	6.69	6.08	5.65	5.33	5.09	4.89	4.73	4.48	4.23	3.96	3.82	3.68	3.53	3.38	3.22	3.05
24	14.03	9.34	7.55	6.59	5.98	5.55	5.23	4.99	4.80	4.64	4.39	4.14	3.87	3.74	3.59	3.45	3.29	3.14	2.97
25	13.88	9.22	7.45	6.49	5.88	5.46	5.15	4.91	4.71	4.56	4.31	4.06	3.79	3.66	3.52	3.37	3.22	3.06	2.89
26	13.74	9.12	7.36	6.41	5.80	5.38	5.07	4.83	4.64	4.48	4.24	3.99	3.72	3.59	3.44	3.30	3.15	2.99	2.82
27	13.61	9.02	7.27	6.33	5.73	5.31	5.00	4.76	4.57	4.41	4.17	3.92	3.66	3.52	3.38	3.23	3.08	2.92	2.75

Table T6 (continued)
UPPER α PROBABILITY POINTS OF F DISTRIBUTION
(ENTRIES ARE $F_{\alpha; \nu_1, \nu_2}$)

Instructions:

(1) Enter the section of the table corresponding to the predetermined α-level.
(2) Enter the row that corresponds to the denominator degrees-of-freedom (ν_2).
(3) Pick the value of F in that row from the column that corresponds to the numerator degrees-of-freedom (ν_1).

ν_2	1	2	3	4	5	6	7	8	9	10	12	15	20	24	30	40	60	120	∞
28	13.50	8.93	7.19	6.25	5.66	5.24	4.93	4.69	4.50	4.35	4.11	3.86	3.60	3.46	3.32	3.18	3.02	2.86	2.69
29	13.39	8.85	7.12	6.19	5.59	5.18	4.87	4.64	4.45	4.29	4.05	3.80	3.54	3.41	3.27	3.12	2.97	2.81	2.64
30	13.29	8.77	7.05	6.12	5.53	5.12	4.82	4.58	4.39	4.24	4.00	3.75	3.49	3.36	3.22	3.07	2.92	2.76	2.59
40	12.61	8.25	6.60	5.70	5.13	4.73	4.44	4.21	4.02	3.87	3.64	3.40	3.15	3.01	2.87	2.73	2.57	2.41	2.23
60	11.97	7.76	6.17	5.31	4.76	4.37	4.09	3.87	3.69	3.54	3.31	3.08	2.83	2.69	2.55	2.41	2.25	2.08	1.89
120	11.38	7.32	5.79	4.95	4.42	4.04	3.77	3.55	3.38	3.24	3.02	2.78	2.53	2.40	2.26	2.11	1.95	1.76	1.54
∞	10.83	6.91	5.42	4.62	4.10	3.74	3.47	3.27	3.10	2.96	2.74	2.51	2.27	2.13	1.99	1.84	1.66	1.45	1.00

ν_1

a   Multiply these entries by 100.

Table T7
TRIANGLE TEST FOR
DIFFERENCE — CRITICAL NUMBER
(MINIMUM) OF CORRECT ANSWERS

Entries are the minimum number of correct responses required for significance at the stated significance level (i.e., column) for the corresponding number of respondents "n" (i.e., row). Reject the assumption of "no difference" if the number of correct responses is greater than or equal to the tabled value.

n	Significance level (%)			
	10	5	1	0.1
3	3	3	—	—
4	4	4	—	—
5	4	4	5	—
6	5	5	6	—
7	5	5	6	7
8	5	6	7	8
9	6	6	7	8
10	6	7	8	9
11	7	7	8	10
12	7	8	9	10
13	8	8	9	11
14	8	9	10	11
15	8	9	10	12
16	9	9	11	12
17	9	10	11	13
18	10	10	12	13
19	10	11	12	14
20	10	11	13	14
21	11	12	13	15
22	11	12	14	15
23	12	12	14	16
24	12	13	15	16
25	12	13	15	17
26	13	14	15	17
27	13	14	16	18
28	14	15	16	18
29	14	15	17	19
30	14	15	17	19
31	15	16	18	20
32	15	16	18	20
33	15	17	18	21
34	16	17	19	21
35	16	17	19	22
36	17	18	20	22
42	19	20	22	25
48	21	22	25	27
54	23	25	27	30
60	26	27	30	33

Table T7 (continued)
TRIANGLE TEST FOR DIFFERENCE — CRITICAL NUMBER (MINIMUM) OF CORRECT ANSWERS

Entries are the minimum number of correct responses required for significance at the stated significance level (i.e., column) for the corresponding number of respondents "n" (i.e., row). Reject the assumption of "no difference" if the number of correct responses is greater than or equal to the tabled value.

	Significance level (%)			
n	10	5	1	0.1
66	28	29	32	35
72	30	32	34	38
78	32	34	37	40
84	35	36	39	43
90	37	38	42	45
96	39	41	44	48

Note: For values of $\underline{n \text{ not}}$ in the table compute $z = (k - (1/3)n)/\sqrt{(2/9)n}$, where k is the number of correct answers. Compare the computed value of z to the critical value of a standard normal random variable (i.e., the values in the last row of Table T4 ($z_\alpha = t_{\alpha,\times}$)).

Table T8
DUO-TRIO TEST FOR DIFFERENCE OR ONE-SIDED PAIRED COMPARISON TEST FOR DIFFERENCE — CRITICAL NUMBER (MINIMUM) OF CORRECT ANSWERS

Entries are the minimum number of correct responses required for significance at the stated significance level (i.e., column) for the corresponding number of respondents "n" (i.e., row). Reject the assumption of "no difference" if the number of correct responses is greater than or equal to the tabled value.

	Significance level (%)			
n	10	5	1	0.1
4	4	—	—	—
5	5	5	—	—
6	6	6	—	—
7	6	7	7	—
8	7	7	8	—
9	7	8	9	—
10	8	9	10	10
11	9	9	10	11
12	9	10	11	12

Table T8 (continued)
DUO-TRIO TEST FOR DIFFERENCE
OR ONE-SIDED PAIRED
COMPARISON TEST FOR
DIFFERENCE — CRITICAL NUMBER
(MINIMUM) OF CORRECT ANSWERS

Entries are the minimum number of correct responses required for significance at the stated significance level (i.e., column) for the corresponding number of respondents "n" (i.e., row). Reject the assumption of "no difference" if the number of correct responses is greater than or equal to the tabled value.

Significance level (%)

n	10	5	1	0.1
13	10	10	12	13
14	10	11	12	13
15	11	12	13	14
16	12	12	14	15
17	12	13	14	16
18	13	13	15	16
19	13	14	15	17
20	14	15	16	18
21	14	15	17	18
22	15	16	17	19
23	16	16	18	20
24	16	17	19	20
25	17	18	19	21
26	17	18	20	22
27	18	19	20	22
28	18	19	21	23
29	19	20	22	24
30	20	20	22	24
31	20	21	23	25
32	21	22	24	26
33	21	22	24	26
34	22	23	25	27
35	22	23	25	27
36	23	24	26	28
40	25	26	28	31
44	27	28	31	33
48	29	31	33	36
52	32	33	35	38
56	34	35	38	40
60	36	37	40	43
64	38	40	42	45
68	40	42	45	48
72	42	44	47	50
76	45	46	49	52
80	47	48	51	55
84	49	51	54	57
88	51	53	56	59
92	53	55	58	62

Table T8 (continued)
DUO-TRIO TEST FOR DIFFERENCE
OR ONE-SIDED PAIRED
COMPARISON TEST FOR
DIFFERENCE — CRITICAL NUMBER
(MINIMUM) OF CORRECT ANSWERS

Entries are the minimum number of correct responses required for significance at the stated significance level (i.e., column) for the corresponding number of respondents "n" (i.e., row). Reject the assumption of "no difference" if the number of correct responses is greater than or equal to the tabled value.

n	Significance level (%)			
	10	5	1	0.1
96	55	57	60	64
100	57	59	63	66

Note: For values of n not in the table compute $z = (k - 0.5n)/\sqrt{0.25n}$, where k is the number of correct answers. Compare the computed value of z to the critical value of a standard normal random variable (i.e., the values in the last row of Table T4 ($z_\alpha = t_{\alpha,\infty}$)).

Table T9
TWO-SIDED PAIRED COMPARISON
TEST FOR DIFFERENCE —
CRITICAL NUMBER (MINIMUM) OF
CORRECT ANSWERS

Entries are the minimum number of correct responses required for significance at the stated significance level (i.e., column) for the corresponding number of respondents "n" (i.e., row). Reject the assumption of "no difference" if the number of correct responses is greater than or equal to the tabled value.

n	Significance level (%)			
	10	5	1	0.1
5	5	—	—	—
6	6	6	—	—
7	7	7	—	—
8	7	8	8	—
9	8	8	9	—
10	9	9	10	—
11	9	10	11	11
12	10	10	11	12
13	10	11	12	13
14	11	12	13	14
15	12	12	13	14
16	12	13	14	15
17	13	13	15	16

Table T9 (continued)
TWO-SIDED PAIRED COMPARISON TEST FOR DIFFERENCE — CRITICAL NUMBER (MINIMUM) OF CORRECT ANSWERS

Entries are the minimum number of correct responses required for significance at the stated significance level (i.e., column) for the corresponding number of respondents "n" (i.e., row). Reject the assumption of "no difference" if the number of correct responses is greater than or equal to the tabled value.

	Significance level (%)			
n	10	5	1	0.1
18	13	14	15	17
19	14	15	16	17
20	15	15	17	18
21	15	16	17	19
22	16	17	18	19
23	16	17	19	20
24	17	18	19	21
25	18	18	20	21
26	18	19	20	22
27	19	20	21	23
28	19	20	22	23
29	20	21	22	24
30	20	21	23	25
31	21	22	24	25
32	22	23	24	26
33	22	23	25	27
34	23	24	25	27
35	23	24	26	28
36	24	25	27	29
40	26	27	29	31
44	28	29	31	34
48	31	32	34	36
52	33	34	36	39
56	35	36	39	41
60	37	39	41	44
64	40	41	43	46
68	42	43	46	48
72	44	45	48	51
76	46	48	50	53
80	48	50	52	56
84	51	52	55	58
88	53	54	57	60
92	55	56	59	63
96	57	59	62	65
100	59	61	64	67

Note: For values of n not in the table compute $z = (k - 0.5n)/\sqrt{0.25n}$, where k is the number of correct answers. Compare the computed value of z to the critical value of a standard normal random variable (i.e., the values in the last row in Table T4 ($z_{\alpha/2} = t_{\alpha/2,\infty}$)).

Table T10
TWO-OUT-OF-FIVE TEST FOR DIFFERENCE — CRITICAL NUMBER (MINIMUM) OF CORRECT ANSWERS

Entries are the minimum number of correct responses required for significance at the stated significance level (i.e., column) for the corresponding number of respondents "n" (i.e., row). Reject the assumption of "no difference" if the number of correct responses is greater than or equal to the tabled value.

	Significance level (%)			
n	10	5	1	0.1
2	2	2	2	—
3	2	2	3	3
4	2	3	3	4
5	2	3	3	4
6	3	3	4	5
7	3	3	4	5
8	3	3	4	5
9	3	4	4	5
10	3	4	5	6
11	3	4	5	6
12	4	4	5	6
13	4	4	5	6
14	4	4	5	7
15	4	5	6	7
16	4	5	6	7
17	4	5	6	7
18	4	5	6	8
19	5	5	6	8
20	5	5	7	8
21	5	6	7	8
22	5	6	7	8
23	5	6	7	9
24	5	6	7	9
25	5	6	7	9
26	6	6	8	9
27	6	6	8	9
28	6	7	8	10
29	6	7	8	10
30	6	7	8	10
31	6	7	8	10
32	6	7	9	10
33	7	7	9	11
34	7	7	9	11
35	7	8	9	11
36	7	8	9	11
37	7	8	9	11
38	7	8	10	11
39	7	8	10	12
40	7	8	10	12

Table T10 (continued)
TWO-OUT-OF-FIVE TEST FOR DIFFERENCE — CRITICAL NUMBER (MINIMUM) OF CORRECT ANSWERS

Entries are the minimum number of correct responses required for significance at the stated significance level (i.e., column) for the corresponding number of respondents "n" (i.e., row). Reject the assumption of "no difference" if the number of correct responses is greater than or equal to the tabled value.

	Significance level (%)			
n	10	5	1	0.1
41	8	8	10	12
42	8	9	10	12
43	8	9	10	12
44	8	9	11	12
45	8	9	11	13
46	8	9	11	13
47	8	9	11	13
48	9	9	11	13
49	9	10	11	13
50	9	10	11	14
51	9	10	12	14
52	9	10	12	14
53	9	10	12	14
54	9	10	12	14
55	9	10	12	14
56	10	10	12	14
57	10	11	12	15
58	10	11	13	15
59	10	11	13	15
60	10	11	13	15
70	11	12	14	17
80	13	14	16	18
90	14	15	17	20
100	15	16	19	21

Note: For values of n not in the table compute $z = (k - 0.1n)/\sqrt{0.09n}$, where k is the number of correct answers. Compare the computed value of z to the critical value of a standard normal random variable (i.e., the values in the last row of Table T4 ($z_\alpha = t_{\alpha,\infty}$)).

Table T11
TRIANGLE TEST FOR SIMILARITY —
CRITICAL NUMBER (MAXIMUM) OF
CORRECT ANSWERS

Accept the null hypothesis of no difference with $100(1 - \beta)\%$ confidence if the number of correct responses is less than or equal to the number in the table that corresponds to the specified values of n, β, and p_d, where p_d is the proportion of the population that can distinguish the samples.

n	β	p_d 0.10	0.15	0.20	0.25	0.30
18	0.001	—	—	—	—	—
	0.01	—	—	—	—	—
	0.05	—	—	—	—	—
	0.10	—	—	—	—	6
21	0.001	—	—	—	—	—
	0.01	—	—	—	—	—
	0.05	—	—	—	—	—
	0.10	—	—	—	7	7
24	0.001	—	—	—	—	—
	0.01	—	—	—	—	—
	0.05	—	—	—	—	8
	0.10	—	—	—	8	9
27	0.001	—	—	—	—	—
	0.01	—	—	—	—	—
	0.05	—	—	—	—	9
	0.10	—	—	—	9	10
30	0.001	—	—	—	—	—
	0.01	—	—	—	—	—
	0.05	—	—	—	10	11
	0.10	—	—	10	10	11
33	0.001	—	—	—	—	—
	0.01	—	—	—	—	—
	0.05	—	—	—	11	12
	0.10	—	—	11	12	13
36	0.001	—	—	—	—	—
	0.01	—	—	—	—	—
	0.05	—	—	—	12	13
	0.10	—	—	12	13	14
39	0.001	—	—	—	—	—
	0.01	—	—	—	—	13
	0.05	—	—	—	13	15
	0.10	—	—	13	15	16
42	0.001	—	—	—	—	—
	0.01	—	—	—	—	14
	0.05	—	—	—	15	16
	0.10	—	—	14	16	17

Table T11 (continued)
TRIANGLE TEST FOR SIMILARITY —
CRITICAL NUMBER (MAXIMUM) OF
CORRECT ANSWERS

Accept the null hypothesis of no difference with $100(1 - \beta)\%$ confidence if the number of correct responses is less than or equal to the number in the table that corresponds to the specified values of n, β, and p_d, where p_d is the proportion of the population that can distinguish the samples.

n	β	p_d 0.10	0.15	0.20	0.25	0.30
45	0.001	—	—	—	—	—
	0.01	—	—	—	—	15
	0.05	—	—	15	16	17
	0.10	—	—	16	17	19
48	0.001	—	—	—	—	—
	0.01	—	—	—	—	17
	0.05	—	—	16	17	19
	0.10	—	—	17	19	20
51	0.001	—	—	—	—	—
	0.01	—	—	—	—	18
	0.05	—	—	17	19	20
	0.10	—	17	18	20	22
54	0.001	—	—	—	—	—
	0.01	—	—	—	18	19
	0.05	—	—	18	20	22
	0.10	—	18	20	21	23
57	0.001	—	—	—	—	—
	0.01	—	—	—	19	21
	0.05	—	—	19	21	23
	0.10	—	19	21	23	25
60	0.001	—	—	—	—	—
	0.01	—	—	—	20	22
	0.05	—	—	21	23	25
	0.10	—	20	22	24	26
66	0.001	—	—	—	—	22
	0.01	—	—	—	23	25
	0.05	—	—	23	25	28
	0.10	—	22	25	27	29
72	0.001	—	—	—	—	24
	0.01	—	—	—	25	28
	0.05	—	—	26	28	30
	0.10	—	25	27	30	32
78	0.001	—	—	—	—	27
	0.01	—	—	—	28	30
	0.05	—	26	28	31	33
	0.10	—	27	30	32	35

Table T11 (continued)
TRIANGLE TEST FOR SIMILARITY — CRITICAL NUMBER (MAXIMUM) OF CORRECT ANSWERS

Accept the null hypothesis of no difference with $100(1 - \beta)\%$ confidence if the number of correct responses is less than or equal to the number in the table that corresponds to the specified values of n, β, and p_d, where p_d is the proportion of the population that can distinguish the samples.

n	β	p_d 0.10	0.15	0.20	0.25	0.30
84	0.001	—	—	—	—	30
	0.01	—	—	28	30	33
	0.05	—	28	31	33	36
	0.10	—	30	32	35	38
90	0.001	—	—	—	—	32
	0.01	—	—	30	33	36
	0.05	—	30	33	36	39
	0.10	—	32	35	38	41
96	0.001	—	—	—	32	35
	0.01	—	—	33	36	39
	0.05	—	33	36	39	42
	0.10	—	34	38	41	44

Note: For values of n not in the table calculate the $100(1 - \beta)\%$ upper one-tailed confidence interval — $(1.5(x/n) - 0.5) + (1.5)z_\beta\sqrt{(nx - x^2)/n^3}$, where x is the number of correct answers from the study, n is the number of respondents, and z_β is the upper-β critical value of a standard normal deviate. It may be concluded with $100(1 - \beta)\%$ confidence that the true proportion of distinguishers in the population is no greater than the calculated value. To find z_β use the last row of Table T4, substituting α for β.

Table T12
DUO-TRIO TEST FOR SIMILARITY OR TWO-SIDED PAIRED-COMPARISON TEST FOR SIMILARITY — CRITICAL NUMBER (MAXIMUM) OF CORRECT ANSWERS

Accept the null hypothesis of no difference with $100(1 - \beta)\%$ confidence if the number of correct responses is less than or equal to the number in the table that corresponds to the specified values of n, β, and p_d, where p_d is the proportion of the population that can distinguish the samples.

n	β	p_d 0.10	0.15	0.20	0.25	0.30
24	0.001	—	—	—	—	—
	0.01	—	—	—	—	—

Table T12 (continued)
DUO-TRIO TEST FOR SIMILARITY OR TWO-SIDED PAIRED-COMPARISON TEST FOR SIMILARITY — CRITICAL NUMBER (MAXIMUM) OF CORRECT ANSWERS

Accept the null hypothesis of no difference with $100(1 - \beta)\%$ confidence if the number of correct responses is less than or equal to the number in the table that corresponds to the specified values of n, β, and p_d, where p_d is the proportion of the population that can distinguish the samples.

n	β	p_d 0.10	0.15	0.20	0.25	0.30
	0.05	—	—	—	—	—
	0.10	—	—	—	—	12
28	0.001	—	—	—	—	—
	0.01	—	—	—	—	—
	0.05	—	—	—	—	—
	0.10	—	—	—	—	14
32	0.001	—	—	—	—	—
	0.01	—	—	—	—	—
	0.05	—	—	—	—	—
	0.10	—	—	—	—	16
36	0.001	—	—	—	—	—
	0.01	—	—	—	—	—
	0.05	—	—	—	—	18
	0.10	—	—	—	18	19
40	0.001	—	—	—	—	—
	0.01	—	—	—	—	—
	0.05	—	—	—	—	20
	0.10	—	—	—	20	21
44	0.001	—	—	—	—	—
	0.01	—	—	—	—	—
	0.05	—	—	—	—	22
	0.10	—	—	—	22	24
48	0.001	—	—	—	—	—
	0.01	—	—	—	—	—
	0.05	—	—	—	—	25
	0.10	—	—	—	25	26
52	0.001	—	—	—	—	—
	0.01	—	—	—	—	—
	0.05	—	—	—	26	27
	0.10	—	—	26	27	28
56	0.001	—	—	—	—	—
	0.01	—	—	—	—	—
	0.05	—	—	—	28	29
	0.10	—	—	28	29	31
60	0.001	—	—	—	—	—
	0.01	—	—	—	—	—
	0.05	—	—	—	30	32
	0.10	—	—	30	32	33

Table T12 (continued)
DUO-TRIO TEST FOR SIMILARITY OR
TWO-SIDED PAIRED-COMPARISON
TEST FOR SIMILARITY — CRITICAL
NUMBER (MAXIMUM) OF CORRECT
ANSWERS

Accept the null hypothesis of no difference with $100(1 - \beta)\%$ confidence if the number of correct responses is less than or equal to the number in the table that corresponds to the specified values of n, β, and p_d, where p_d is the proportion of the population that can distinguish the samples.

				p_d		
n	β	0.10	0.15	0.20	0.25	0.30
64	0.001	—	—	—	—	—
	0.01	—	—	—	—	32
	0.05	—	—	—	33	34
	0.10	—	—	32	34	36
68	0.001	—	—	—	—	—
	0.01	—	—	—	—	34
	0.05	—	—	—	35	37
	0.10	—	—	35	36	38
72	0.001	—	—	—	—	—
	0.01	—	—	—	—	36
	0.05	—	—	—	37	39
	0.10	—	—	37	39	41
76	0.001	—	—	—	—	—
	0.01	—	—	—	—	39
	0.05	—	—	38	40	41
	0.10	—	—	39	41	43
80	0.001	—	—	—	—	—
	0.01	—	—	—	—	41
	0.05	—	—	40	42	44
	0.10	—	—	41	43	46
84	0.001	—	—	—	—	—
	0.01	—	—	—	—	43
	0.05	—	—	42	44	46
	0.10	—	—	44	46	48
88	0.001	—	—	—	—	—
	0.01	—	—	—	—	46
	0.05	—	—	44	46	49
	0.10	—	44	46	48	50
92	0.001	—	—	—	—	—
	0.01	—	—	—	46	48
	0.05	—	—	46	49	51
	0.10	—	46	48	51	53
96	0.001	—	—	—	—	—
	0.01	—	—	—	48	50

Table T12 (continued)
DUO-TRIO TEST FOR SIMILARITY OR TWO-SIDED PAIRED-COMPARISON TEST FOR SIMILARITY — CRITICAL NUMBER (MAXIMUM) OF CORRECT ANSWERS

Accept the null hypothesis of no difference with $100(1 - \beta)\%$ confidence if the number of correct responses is less than or equal to the number in the table that corresponds to the specified values of n, β, and p_d, where p_d is the proportion of the population that can distinguish the samples.

n	β	p_d				
		0.10	**0.15**	**0.20**	**0.25**	**0.30**
	0.05	—	—	49	51	54
	0.10	—	48	50	53	55

Note: For values of n not in the table calculate the $100(1 - \beta)\%$ upper one-tailed confidence interval — $(2(x/n) - 1) + (2)z_\beta\sqrt{(nx - x^2)/n^3}$, where x is the number of correct answers from the study, n is the number of respondents, and z_β is the upper-β critical value of a standard normal deviate. It may be concluded with $100(1 - \beta)\%$ confidence that the true proportion of distinguishers in the population is no greater than the calculated value. To find z_β use the last row of Table T4, substituting α for β.

Table T13
TWO-OUT-OF-FIVE TEST FOR SIMILARITY — CRITICAL NUMBER OF CORRECT RESPONSES

Accept the null hypothesis of no difference with $100(1 - \beta)\%$ confidence if the number of correct responses is less than or equal to the number in the table that corresponds to the specified values of n, β, and p_d, where p_d is the proportion of the population that can distinguish the samples.

n	β	p_d				
		0.10	**0.15**	**0.20**	**0.25**	**0.30**
15	0.001	—	—	—	—	—
	0.01	—	—	—	—	—
	0.05	—	—	—	—	2
	0.10	—	—	—	2	2
20	0.001	—	—	—	—	—
	0.01	—	—	—	—	2
	0.05	—	—	—	2	3
	0.10	—	—	2	3	4
25	0.001	—	—	—	—	—
	0.01	—	—	—	—	3

Table T13 (continued)
TWO-OUT-OF-FIVE TEST FOR
SIMILARITY — CRITICAL NUMBER OF
CORRECT RESPONSES

Accept the null hypothesis of no difference with $100(1 - \beta)\%$ confidence if the number of correct responses is less than or equal to the number in the table that corresponds to the specified values of n, β, and p_d, where p_d is the proportion of the population that can distinguish the samples.

n	β	p_d				
		0.10	0.15	0.20	0.25	0.30
	0.05	—	—	—	3	4
	0.10	—	—	3	4	5
30	0.001	—	—	—	—	3
	0.01	—	—	—	3	4
	0.05	—	—	4	5	6
	0.10	—	3	4	6	7
35	0.001	—	—	—	—	4
	0.01	—	—	—	4	6
	0.05	—	—	5	6	7
	0.10	—	4	5	7	8
40	0.001	—	—	—	4	5
	0.01	—	—	4	5	7
	0.05	—	4	6	7	9
	0.10	4	5	7	8	10
45	0.001	—	—	—	5	6
	0.01	—	—	5	7	8
	0.05	—	5	7	9	10
	0.10	—	6	8	10	12
50	0.001	—	—	—	6	7
	0.01	—	—	6	8	10
	0.05	—	6	8	10	12
	0.10	5	7	9	11	13
55	0.001	—	—	—	7	9
	0.01	—	—	7	9	11
	0.05	—	7	9	11	14
	0.10	6	8	10	12	15
60	0.001	—	—	6	8	10
	0.01	—	6	8	10	13
	0.05	6	8	10	13	15
	0.10	7	9	11	14	16
65	0.001	—	—	7	9	12
	0.01	—	7	9	12	14
	0.05	—	9	11	14	17
	0.10	7	10	13	15	18
70	0.001	—	—	8	10	13
	0.01	—	8	10	13	16

Table T13 (continued)
TWO-OUT-OF-FIVE TEST FOR SIMILARITY — CRITICAL NUMBER OF CORRECT RESPONSES

Accept the null hypothesis of no difference with $100(1 - \beta)\%$ confidence if the number of correct responses is less than or equal to the number in the table that corresponds to the specified values of n, β, and p_d, where p_d is the proportion of the population that can distinguish the samples.

n	β	p_d 0.10	0.15	0.20	0.25	0.30
	0.05	7	10	13	15	18
	0.10	8	11	14	17	20
75	0.001	—	—	9	11	14
	0.01	—	9	11	14	17
	0.05	8	11	14	17	20
	0.10	9	12	15	18	21
80	0.001	—	—	10	13	17
	0.01	—	9	12	16	19
	0.05	9	12	15	18	22
	0.10	10	13	16	20	23
85	0.001	—	—	11	14	17
	0.01	—	10	14	17	20
	0.05	9	13	16	20	23
	0.10	11	14	18	21	25
90	0.001	—	9	12	15	19
	0.01	—	11	15	18	22
	0.05	10	14	17	21	25
	0.10	11	15	19	23	26
95	0.001	—	—	13	16	20
	0.01	—	12	16	20	23
	0.05	11	15	19	22	26
	0.10	12	16	20	24	28
100	0.001	—	10	14	18	22
	0.01	—	13	17	21	25
	0.05	12	16	20	24	28
	0.10	13	17	21	26	30

Note: For values of n not in the table calculate the $100(1 - \beta)\%$ upper one-tailed confidence interval — $((10/9)(x/n) - (1/9)) + (10.9)z_\beta \sqrt{(nx - x^2)/n^3}$, where x is the number of correct answers from the study, n is the number of respondents, and z_β is the upper-β critical value of a standard normal deviate. It may be concluded with $100(1 - \beta)\%$ confidence that the true proportion of distinguishers in the population is no greater than the calculated value. To find z_β use the last row of Table T4, substituting α for β.

Table T14
PERCENTAGE POINTS OF THE STUDENTIZED RANGE — UPPER α CRITICAL VALUES FOR TUKEY'S HSD MULTIPLE COMPARISON PROCEDURE

Instructions:

(1) Enter the section of the table that corresponds to the predetermined α-level.

(2) Enter the row that corresponds to the degrees-of-freedom for error from the ANOVA.

(3) Pick the value of q in that row from the column that corresponds to the number of treatments being compared.

t

ν	2	3	4	5	6	7	8	9	10

The Entries are $q_{0.01}$, where $P(q < q_{0.01}) = 0.99$

ν	2	3	4	5	6	7	8	9	10
1	90.03	135.0	164.3	185.6	202.2	215.8	227.2	237.0	245.6
2	14.04	19.02	22.29	24.72	26.63	28.20	29.53	30.68	31.69
3	8.26	10.62	12.17	13.33	14.24	15.00	15.64	16.20	16.69
4	6.51	8.12	9.17	9.96	10.58	11.10	11.55	11.93	12.27
5	5.70	6.98	7.80	8.42	8.91	9.32	9.67	9.97	10.24
6	5.24	6.33	7.03	7.56	7.97	8.32	8.61	8.87	9.10
7	4.95	5.92	6.54	7.01	7.37	7.68	7.94	8.17	8.37
8	4.75	5.64	6.20	6.62	6.96	7.24	7.47	7.68	7.86
9	4.60	5.43	5.96	6.35	6.66	6.91	7.13	7.33	7.49
10	4.48	5.27	5.77	6.14	6.43	6.67	6.87	7.05	7 21
11	4.39	5.15	5.62	5.97	6.25	6.48	6.67	6.84	6.99
12	4.32	5.05	5.50	5.84	6.10	6.32	6.51	6.67	6.81
13	4.26	4.96	5.40	5.73	5.98	6.19	6.37	6.53	6.67
14	4.21	4.89	5.32	5.63	5.88	6.08	6.26	6.41	6.54
15	4.17	4.84	5.25	5.56	5.80	5.99	6.16	6.31	6.44
16	4.13	4.79	5.19	5.49	5.72	5.92	6.08	6.22	6.35
17	4.10	4.74	5.14	5.43	5.66	5.85	6.01	6.15	6.27
18	4.07	4.70	5.09	5.38	5.60	5.79	5.94	6.08	6.20
19	4.05	4.67	5.05	5.33	5.55	5.73	5.89	6.02	6.14
20	4.02	4.64	5.02	5.29	5.51	5.69	5.84	5.97	6.09
24	3.96	4.55	4.91	5.17	5.37	5.54	5.69	5.81	5.92
30	3.89	4.45	4.80	5.05	5.24	5.40	5.54	5.65	5.76
40	3.82	4.37	4.70	4.93	5.11	5.26	5.39	5.50	5.60
60	3.76	4.28	4.59	4.82	4.99	5.13	5.25	5.36	5.45
120	3.70	4.20	4.50	4.71	4.87	5.01	5.12	5.21	5.30
∞	3.64	4.12	4.40	4.60	4.76	4.88	4.99	5.08	5.16

t

ν	11	12	13	14	15	16	17	18	19	20
1	253.2	260.0	266.2	271.8	277.0	281.8	286.3	290.4	294.3	298.0
2	32.59	33.40	34.13	34.81	35.43	36.00	36.53	37.03	37.50	37.95
3	17.13	17.53	17.89	18.22	18.52	18.81	19.07	19.32	19.55	19.77
4	12.57	12.84	13.09	13.32	13.53	13.73	13.91	14.08	14.24	14.40
5	10.48	10.70	10.89	11.08	11.24	11.40	11.55	11.68	11.81	11.93
6	9.30	9.48	9.65	9.81	9.95	10.08	10.21	10.32	10.43	10.54
7	8.55	8.71	8.86	9.00	9.12	9.24	9.35	9.46	9.55	9.65
8	8.03	8.18	8.31	8.44	8.55	8.66	8.76	8.85	8.94	9.03

Table T14 (continued)
PERCENTAGE POINTS OF THE STUDENTIZED RANGE — UPPER α
CRITICAL VALUES FOR TUKEY'S HSD MULTIPLE COMPARISON PROCEDURE

Instructions:
(1) Enter the section of the table that corresponds to the predetermined α-level.
(2) Enter the row that corresponds to the degrees-of-freedom for error from the ANOVA.
(3) Pick the value of q in that row from the column that corresponds to the number of treatments being compared.

ν	11	12	13	14	15	16	17	18	19	20
9	7.65	7.78	7.91	8.03	8.13	8.23	8.33	8.41	8.49	8.57
10	7.36	7.49	7.60	7.71	7.81	7.91	7.99	8.08	8.15	8.23
11	7.13	7.25	7.36	7.46	7.56	7.65	7.73	7.81	7.88	7.95
12	6.94	7.06	7.17	7.26	7.36	7.44	7.52	7.59	7.66	7.73
13	6.79	6.90	7.01	7.10	7.19	7.27	7.35	7.42	7.48	7.55
14	6.66	6.77	6.87	6.96	7.05	7.13	7.20	7.27	7.33	7.39
15	6.55	6.66	6.76	6.84	6.93	7.00	7.07	7.14	7.20	7.26
16	6.46	6.56	6.66	6.74	6.82	6.90	6.97	7.03	7.09	7.15
17	6.38	6.48	6.57	6.66	6.73	6.81	6.87	6.94	7.00	7.05
18	6.31	6.41	6.50	6.58	6.65	6.73	6.79	6.85	6.91	6.97
19	6.25	6.34	6.43	6.51	6.58	6.65	6.72	6.78	6.84	6.89
20	6.19	6.28	6.37	6.45	6.52	6.59	6.65	6.71	6.77	6.82
24	6.02	6.11	6.19	6.26	6.33	6.39	6.45	6.51	6.56	6.61
30	5.85	5.93	6.01	6.08	6.14	6.20	6.26	6.31	6.36	6.41
40	5.69	5.76	5.83	5.90	5.96	6.02	6.07	6.12	6.16	6.21
60	5.53	5.60	5.67	5.73	5.78	5.84	5.89	5.93	5.97	6.01
120	5.37	5.44	5.50	5.56	5.61	5.66	5.71	5.75	5.79	5.83
∞	5.23	5.29	5.35	5.40	5.45	5.49	5.54	5.57	5.61	5.65

The Entries are $q_{0.05}$ where $P(q < q_{0.05}) = 0.95$

ν	2	3	4	5	6	7	8	9	10
1	17.97	26.98	32.82	37.08	40.41	43.12	45.40	47.36	49.07
2	6.08	8.33	9.80	10.88	11.74	12.44	13.03	13.54	13.99
3	4.50	5.91	6.82	7.50	8.04	8.48	8.85	9.18	9.46
4	3.93	5.04	5.76	6.29	6.71	7.05	7.35	7.60	7.83
5	3.64	4.60	5.22	5.67	6.03	6.33	6.58	6.80	6.99
6	3.46	4.34	4.90	5.30	5.63	5.90	6.12	6.32	6.49
7	3.34	4.16	4.68	5.06	5.36	5.61	5.82	6.00	6.16
8	3.26	4.04	4.53	4.89	5.17	5.40	5.60	5.77	5.92
9	3.20	3.95	4.41	4.76	5.02	5.24	5.43	5.59	5.74
10	3.15	3.88	4.33	4.65	4.91	5.12	5.30	5.46	5.60
11	3.11	3.82	4.26	4.57	4.82	5.03	5.20	5.35	5.49
12	3.08	3.77	4.20	4.51	4.75	4.95	5.12	5.27	5.39
13	3.06	3.73	4.15	4.45	4.69	4.88	5.05	5.19	5.32

Table T14 (continued)
PERCENTAGE POINTS OF THE STUDENTIZED RANGE — UPPER α CRITICAL VALUES FOR TUKEY'S HSD MULTIPLE COMPARISON PROCEDURE

Instructions:

(1) Enter the section of the table that corresponds to the predetermined α-level.

(2) Enter the row that corresponds to the degrees-of-freedom for error from the ANOVA.

(3) Pick the value of q in that row from the column that corresponds to the number of treatments being compared.

The Entries are $q_{0.05}$ where $P(q < q_{0.05}) = 0.95$

ν	2	3	4	5	6	7	8	9	10
14	3.03	3.70	4.11	4.41	4.64	4.83	4.99	5.13	5.25
15	3.01	3.67	4.08	4.37	4.59	4.78	4.94	5.08	5.20
16	3.00	3.65	4.05	4.33	4.56	4.74	4.90	5.03	5.15
17	2.98	3.63	4.02	4.30	4.52	4.70	4.86	4.99	5.11
18	2.97	3.61	4.00	4.28	4.49	4.67	4.82	4.96	5.07
19	2.96	3.59	3.98	4.25	4.47	4.65	4.79	4.92	5.04
20	2.95	3.58	3.96	4.23	4.45	4.62	4.77	4.90	5.01
24	2.92	3.53	3.90	4.17	4.37	4.54	4.68	4.81	4.92
30	2.89	3.49	3.85	4.10	4.30	4.46	4.60	4.72	4.82
40	2.86	3.44	3.79	4.04	4.23	4.39	4.52	4.63	4.73
60	2.83	3.40	3.74	3.98	4.16	4.31	4.44	4.55	4.65
120	2.80	3.36	3.68	3.92	4.10	4.24	4.36	4.47	4.56
∞	2.77	3.31	3.63	3.86	4.03	4.17	4.29	4.39	4.47

ν	11	12	13	14	15	16	17	18	19	20
1	50.59	51.96	53.20	54.33	55.36	56.32	57.22	58.04	58.83	59.56
2	14.39	14.75	15.08	15.38	15.65	15.91	16.14	16.37	16.57	16.77
3	9.72	9.95	10.15	10.35	10.53	10.69	10.84	10.98	11.11	11.24
4	8.03	8.21	8.37	8.52	8.66	8.79	8.91	9.03	9.13	9.23
5	7.17	7.32	7.47	7.60	7.72	7.83	7.93	8.03	8.12	8.21
6	6.65	6.79	6.92	7.03	7.14	7.24	7.34	7.43	7.51	7.59
7	6.30	6.43	6.55	6.66	6.76	6.85	6.94	7.02	7.10	7.17
8	6.05	6.18	6.29	6.39	6.48	6.57	6.65	6.73	6.80	6.87
9	5.87	5.98	6.09	6.19	6.28	6.36	6.44	6.51	6.58	6.64
10	5.72	5.83	5.93	6.03	6.11	6.19	6.27	6.34	6.40	6.47
11	5.61	5.71	5.81	5.90	5.98	6.06	6.13	6.20	6.27	6.33
12	5.51	5.61	5.71	5.80	5.88	5.95	6.02	6.09	6.15	6.21
13	5.43	5.53	5.63	5.71	5.79	5.86	5.93	5.99	6.05	6.11
14	5.36	5.46	5.55	5.64	5.71	5.79	5.85	5.91	5.97	6.03
15	5.31	5.40	5.49	5.57	5.65	5.72	5.78	5.85	5.90	5.96
16	5.26	5.35	5.44	5.52	5.59	5.66	5.73	5.79	5.84	5.90
17	5.21	5.31	5.39	5.47	5.54	5.61	5.67	5.73	5.79	5.84
18	5.17	5.27	5.35	5.43	5.50	5.57	5.63	5.69	5.74	5.79
19	5.14	5.23	5.31	5.39	5.46	5.53	5.59	5.65	5.70	5.75
20	5.11	5.20	5.28	5.36	5.43	5.49	5.55	5.61	5.66	5.71

Table T14 (continued)
PERCENTAGE POINTS OF THE STUDENTIZED RANGE — UPPER α CRITICAL VALUES FOR TUKEY'S HSD MULTIPLE COMPARISON PROCEDURE

Instructions:

(1) Enter the section of the table that corresponds to the predetermined α-level.

(2) Enter the row that corresponds to the degrees-of-freedom for error from the ANOVA.

(3) Pick the value of q in that row from the column that corresponds to the number of treatments being compared.

					t					
ν	11	12	13	14	15	16	17	18	19	20
24	5.01	5.10	5.18	5.25	5.32	5.38	5.44	5.49	5.55	5.59
30	4.92	5.00	5.08	5.15	5.21	5.27	5.33	5.38	5.43	5.47
40	4.82	4.90	4.98	5.04	5.11	5.16	5.22	5.27	5.31	5.36
60	4.73	4.81	4.88	4.94	5.00	5.06	5.11	5.15	5.20	5.24
120	4.64	4.71	4.78	4.84	4.90	4.95	5.00	5.04	5.09	5.13
∞	4.55	4.62	4.68	4.74	4.80	4.85	4.89	4.93	4.97	5.01

The entries are $q_{0.10}$ where $P(q < q_{0.10}) = 0.90$

					t				
ν	2	3	4	5	6	7	8	9	10
1	8.93	13.44	16.36	18.49	20.15	21.51	22.64	23.62	24.48
2	4.13	5.73	6.77	7.54	8.14	8.63	9.05	9.41	9.72
3	3.33	4.47	5.20	5.74	6.16	6.51	6.81	7.06	7.29
4	3.01	3.98	4.59	5.03	5.39	5.68	5.93	6.14	6.33
5	2.85	3.72	4.26	4.66	4.98	5.24	5.46	5.65	5.82
6	2.75	3.56	4.07	4.44	4.73	4.97	5.17	5.34	5.50
7	2.68	3.45	3.93	4.28	4.55	4.78	4.97	5.14	5.28
8	2.63	3.37	3.83	4.17	4.43	4.65	4.83	4.99	5.13
9	2.59	3.32	3.76	4.08	4.34	4.54	4.72	4.87	5.01
10	2.56	3.27	3.70	4.02	4.26	4.47	4.64	4.78	4.91
11	2.54	3.23	3.66	3.96	4.20	4.40	4.57	4.71	4.84
12	2.52	3.20	3.62	3.92	4.16	4.35	4.51	4.65	4.78
13	2.50	3.18	3.59	3.88	4.12	4.30	4.46	4.60	4.72
14	2.49	3.16	3.56	3.85	4.08	4.27	4.42	4.56	4.68
15	2.48	3.14	3.54	3.83	4.05	4.23	4.39	4.52	4.64
16	2.47	3.12	3.52	3.80	4.03	4.21	4.36	4.49	4.61
17	2.46	3.11	3.50	3.78	4.00	4.18	4.33	4.46	4.58
18	2.45	3.10	3.49	3.77	3.98	4.16	4.31	4.44	4.55
19	2.45	3.09	3.47	3.75	3.97	4.14	4.29	4.42	4.53
20	2.44	3.08	3.46	3.74	3.95	4.12	4.27	4.40	4.51
24	2.42	3.05	3.42	3.69	3.90	4.07	4.21	4.34	4.44
30	2.40	3.02	3.39	3.65	3.85	4.02	4.16	4.28	4.38
40	2.38	2.99	3.35	3.60	3.80	3.96	4.10	4.21	4.32
60	2.36	2.96	3.31	3.56	3.75	3.91	4.04	4.16	4.25
120	2.34	2.93	3.28	3.52	3.71	3.86	3.99	4.10	4.19

Table T14 (continued)
PERCENTAGE POINTS OF THE STUDENTIZED RANGE — UPPER α CRITICAL VALUES FOR TUKEY'S HSD MULTIPLE COMPARISON PROCEDURE

Instructions: (1) Enter the section of the table that corresponds to the predetermined α-level.
(2) Enter the row that corresponds to the degrees-of-freedom for error from the ANOVA.
(3) Pick the value of q in that row from the column that corresponds to the number of treatments being compared.

The entries are $q_{0.10}$ where $P(q < q_{0.10}) = 0.90$

t

ν	2	3	4	5	6	7	8	9	10
∞	2.33	2.90	3.24	3.48	3.66	3.81	3.93	4.04	4.13

t

t	11	12	13	14	15	16	17	18	19	20
1	25.24	25.92	26.54	27.10	27.62	28.10	28.54	28.96	29.35	29.71
2	10.01	10.26	10.49	10.70	10.89	11.07	11.24	11.39	11.54	11.68
3	7.49	7.67	7.83	7.98	8.12	8.25	8.37	8.48	8.58	8.68
4	6.49	6.65	6.78	6.91	7.02	7.13	7.23	7.33	7.41	7.50
5	5.97	6.10	6.22	6.34	6.44	6.54	6.63	6.71	6.79	6.86
6	5.64	5.76	5.87	5.98	6.07	6.16	6.25	6.32	6.40	6.47
7	5.41	5.53	5.64	5.74	5.83	5.91	5.99	6.06	6.13	6.19
8	5.25	5.36	5.46	5.56	5.64	5.72	5.80	5.87	5.93	6.00
9	5.13	5.23	5.33	5.42	5.51	5.58	5.66	5.72	5.79	5.85
10	5.03	5.13	5.23	5.32	5.40	5.47	5.54	5.61	5.67	5.73
11	4.95	5.05	5.15	5.23	5.31	5.38	5.45	5.51	5.57	5.63
12	4.89	4.99	5.08	5.16	5.24	5.31	5.37	5.44	5.49	5.55
13	4.83	4.93	5.02	5.10	5.18	5.25	5.31	5.37	5.43	5.48
14	4.79	4.88	4.97	5.05	5.12	5.19	5.26	5.32	5.37	5.43
15	4.75	4.84	4.93	5.01	5.08	5.15	5.21	5.27	5.32	5.38
16	4.71	4.81	4.89	4.97	5.04	5.11	5.17	5.23	5.28	5.33
17	4.68	4.77	4.86	4.93	5.01	5.07	5.13	5.19	5.24	5.30
18	4.65	4.75	4.83	4.90	4.98	5.04	5.10	5.16	5.21	5.26
19	4.63	4.72	4.80	4.88	4.95	5.01	5.07	5.13	5.18	5.23
20	4.61	4.70	4.78	4.85	4.92	4.99	5.05	5.10	5.16	5.20
24	4.54	4.63	4.71	4.78	4.85	4.91	4.97	5.02	5.07	5.12
30	4.47	4.56	4.64	4.71	4.77	4.83	4.89	4.94	4.99	5.03
40	4.41	4.49	4.56	4.63	4.69	4.75	4.81	4.86	4.90	4.95
60	4.34	4.42	4.49	4.56	4.62	4.67	4.73	4.78	4.82	4.86
120	4.28	4.35	4.42	4.48	4.54	4.60	4.65	4.69	4.74	4.78
∞	4.21	4.28	4.35	4.41	4.47	4.52	4.57	4.61	4.65	4.69

INDEX

A

B

C